Mechanism Design and Analysis

Using Creo Mechanism 6.0

Kuang-Hua Chang, Ph.D.
School of Aerospace and Mechanical Engineering
The University of Oklahoma
Norman, OK

Publications

SDC Publications
P.O. Box 1334
Mission, KS 66222
913-262-2664
www.SDCpublications.com
Publisher: Stephen Schroff

ISBN-13: 978-1-63057-298-3
ISBN-10: 1-63057-298-5

Printed and bound in the United States of America.

Preface

This book is prepared to help readers become familiar with *Creo Mechanism*, a module of the *PTC Creo Parametric* software family, which supports modeling and analysis (or simulation) of mechanisms in a virtual (computer) environment. Capabilities in *Creo Mechanism* or simply *Mechanism* allow users to simulate and visualize mechanism performance. Using *Mechanism* early in the product development stage would prevent costly redesign due to design defects found in the prototyping and physical testing phases, therefore contributing to a more cost effective, reliable, and efficient product development process.

This book covers major concepts and frequently used commands required to advance readers from a novice to an intermediate level in using *Mechanism*. Basic concepts discussed in this book include motion model creation, such as body and joint definitions; analysis type selection, including static, position, kinematic and dynamic analyses; and results visualization using graphs, data, and animations. The concept and steps are introduced using simple, yet realistic examples.

Verifying results obtained from computer simulations is extremely important. One of the unique features about this book is the incorporation of theoretical discussions for kinematic and dynamic analyses in conjunction with the simulation results obtained from using *Mechanism*. The purpose of the theoretical discussions lies in simply supporting the verification of simulation results, rather than providing a thorough and in-depth discussion on the subjects of kinematics and dynamics. *Mechanism* is not foolproof. It requires a certain level of experience and expertise to master the software. Before arriving at that level, it will be very helpful for you to verify the simulation results whenever possible. Verifying simulation results will increase your confidence in using the software and prevent you from being fooled (hopefully, only occasionally) by any erroneous simulations produced by the software. Example model files have been prepared for you to go through the lessons. In addition, *Excel* spreadsheets that support the theoretical verifications of selected examples are available. You may download all model files and *Excel* spreadsheets from the web site of *SDC Publications* at:

http://www.sdcpublications.com/

In addition to the files prepared for going over individual lessons, completely assembled models with simulation results are provided for your references. You may want to start each lesson by reviewing the introduction section and opening the assembled model in *Mechanism* to see the motion simulation, in hope of gaining more understanding about the example problems.

This book is written following the project-based learning approach and is intentionally kept simple. Therefore, this book may not contain every single detail about *Mechanism*. For a complete reference on *Mechanism*, you may use on-line help, or visit the web site of *Parametric Technology Corporation* at:

http://www.ptc.com/

This book serves well for instruction of regular classes. It would likely be used as a supplemental textbook for courses like *Mechanism*, *Rigid Body Dynamics*, *Computer-Aided Design*, or *Computer-Aided Engineering*. This book should cover 4 to 6 weeks of class instruction, depending on how the courses are taught and the technical background of the students. Some of the exercise problems given at the end of the lessons may take significant effort for students to complete. The author strongly encourages instructors and/or teaching assistants to go through those exercises before assigning them to students.

This book should also serve self-learners well. If such describes you, you are expected to have basic knowledge in *Physics* and *Mathematics*, preferably a bachelor's or associate degree in science or engineering. In addition, this book assumes that you are familiar with the basic concept and operation of

Creo part and assembly modes. A self-learner should be able to complete all lessons in this book in about 40 hours. An investment of 40 hours should advance you from a novice to an intermediate level user.

KHC
Norman, Oklahoma
June 3, 2019

Acknowledgments

Acknowledgment is due to Mr. Stephen Schroff at *SDC Publications* for his encouragement and help. Without his encouragement, this book would still be in its primitive stage. The contribution in editing, formatting, and publishing the manuscript made by the staff of *SDC Publications* is acknowledged and highly appreciated.

Thanks are also due to undergraduate students at the University of Oklahoma (OU) for their help in testing the examples included in this book. They made numerous suggestions that improved clarity of presentation and found numerous errors that would have otherwise crept into the book. Their contributions to this book are greatly appreciated.

I am grateful to my current and former students, Thomas Cates, Petr Sramek, Dave Oubre, and Trey Wheeler, for their excellent efforts in creating examples for the application lessons, especially *Lessons 9* and *10*. Both the assistive device and the racecar projects were successful and well recognized.

About the Cover Page

The picture shown on the cover page is the solid model of a Formula SAE (Society of Automotive Engineers) style racecar designed and built by engineering students at the University of Oklahoma (OU) during 2005-2006. The racecar model was built in *Creo* with about 1,400 parts and assemblies. Even though this was a team effort, most parts and assemblies were created and managed by then-Senior Mechanical Engineering student, Mr. Dave Oubre (Super Dave). His dedication in creating such a detailed and accurate racecar solid model is admirable. His effort is highly appreciated.

Each year engineering students throughout the world design and build formula-style racecars and participate in the annual Formula SAE competitions. The result is a great experience for young engineers in a meaningful engineering project as well as the opportunity to work in a dedicated team effort. The OU team has been very competitive in the Formula SAE competitions. The team won numerous awards throughout the years and finished 12th and 8th overall at the Formula SAE and Formula SAE West competitions, respectively, in 2006. Their 2005 racecar design also won the prestigious 2005 PTC Award in the Education, Colleges, and Universities category. The worldwide competition is sponsored by Parametric Technology Corporation.

A quarter of the racecar suspension has been employed as the final application example to be discussed in *Lesson 10* of this book. You will find more details of the racecar suspension in that lesson.

Table of Contents

Lesson 1: Introduction to *Mechanism*

Lesson 2: A Ball Throwing Example

Lesson 3: A Spring-Mass System

Lesson 4: A Simple Pendulum

Lesson 5: A Slider-Crank Mechanism

Lesson 6: A Compound Spur Gear Train

Lesson 7: Planetary Gear Train Systems

Lesson 8: Cam and Follower

Lesson 9: Assistive Device for Wheelchair Soccer Game

Lesson 10: Kinematic Analysis for a Racecar Suspension

Appendix A: Defining Joints

Appendix B: Defining Measures

Appendix C: The Default Unit System

Appendix D: Functions

Lesson 1: Introduction to *Mechanism*

1.1 Overview of the Lesson

The purpose of this lesson is to provide you with a brief overview on the *PTC Creo Mechanism* software tool or simply *Mechanism*. *Mechanism* is a virtual prototyping tool that supports mechanism design and analysis, which is also called motion analysis in this book. Instead of building and testing physical prototypes of the mechanism, you may use the *Mechanism* software tool to evaluate and refine the mechanism before finalizing the design and entering the functional prototyping stage. *Mechanism* will help you analyze and eventually design better engineering products. More specifically, the software enables you to size motors and actuators, determine power consumption, layout linkages, develop cams, understand gear trains, size springs and dampers, and determine interference between parts, which would usually require tests of physical prototypes. With such information, you will gain insight on how the mechanism works and why it behaves in certain ways. You will be able to modify the design and often achieve better design alternatives using the more convenient and less expensive virtual prototypes. In the long run, using virtual prototyping tools, such as *Mechanism*, will help you become a more experienced and competent design engineer.

In this lesson, we start with a brief introduction to *Mechanism* and the types of physical problems that *Mechanism* is capable of solving. We will then discuss capabilities offered by *Mechanism* for creating motion models, conducting motion analyses, and viewing motion analysis results. In the last section, we will mention examples employed in this book and major topics to be discussed in these examples.

Note that materials presented in this lesson will be brief. More details on various aspects of mechanism design and analysis using *Mechanism* will be given in later lessons.

1.2 What is *Mechanism*?

Mechanism is a computer software tool that enables engineers to analyze and design mechanisms. *Mechanism* is a module of the *PTC Creo* product family developed by *Parametric Technology Corporation*. This software allows users to create virtual mechanisms that answer general questions in product design such as those described next. An internal combustion engine shown in Figures 1-1 and 1-2 is used to illustrate some of the typical questions in mechanism design.

1. Will the components of the mechanism collide in operation? For example, will the connecting rod collide with the inner surface of the piston or the inner surface of the engine case during operation?

2. Will the components in the mechanism you design move according to your intent? For example, will the piston stay entirely in the piston sleeve? Will the system lock up when the firing force aligns vertically with the connecting rod?

3.   How fast will the components move, e.g., the vertical motion of the piston?

4.   How much torque or force does it take to drive the mechanism? For example, what will be the minimum firing load to move the piston? Note that in this case, proper friction forces must be added to simulate the resistance of the mechanism before a realistic firing force can be calculated.

5.   What is the reaction force or torque generated at a connection (also called *joint* or *constraint*) between components (or bodies) during motion? For example, what is the reaction force at the joint between the connecting rod and the piston pin? This reaction force is critical since the structural integrity of the piston pin and the connecting rod must be ensured; i.e., they must be strong and durable enough to sustain the firing load in operation.

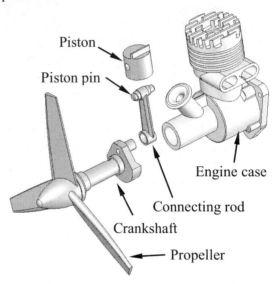

Figure 1-1 Internal Combustion
                Engine (Unexploded View)   Figure 1-2 Internal Combustion Engine (Exploded View)

The modeling and analysis capabilities in *Mechanism* may help you answer these common questions accurately and realistically, as long as the motion model is properly defined. The capabilities available in *Mechanism* also help you search for better design alternatives. A better design alternative is very much problem dependent. It is critical that a design problem be clearly defined by the designer up front before searching for better design alternatives. For the engine example, a better design alternative can be a design that reveals:

1.   A smaller reaction force applied to the connecting rod, and
2.   No collisions or interference between components in motion.

In order to vary component sizes for exploring better design alternatives, the parts and assembly must be adequately parameterized to capture design intents. At the parts level, design parameterization implies creating solid features and relating dimensions properly. At the assembly level, design parameterization involves defining assembly constraints (called placement constraints in *Creo*) and relating dimensions within a single part or across parts. When a solid model is fully parameterized, a change in dimension value can be automatically propagated to all parts affected. Parts affected must be rebuilt successfully, and at the same time, they will have to maintain proper position and orientation with respect to one another without violating any assembly constraints or revealing part penetration or excessive gaps. For example, in this engine example, a change in the bore diameter of the engine case will alter not only the geometry of the case itself, but all other parts affected, such as the piston, piston sleeve, and even the crankshaft, as illustrated in Figure 1-3. Moreover, they all have to be rebuilt properly and the entire assembly must stay intact through assembly constraints.

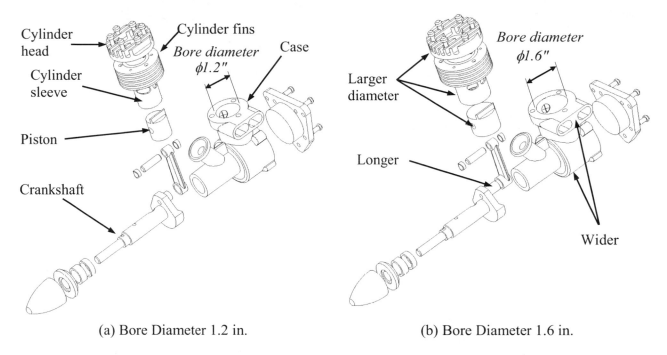

(a) Bore Diameter 1.2 in.   (b) Bore Diameter 1.6 in.

Figure 1-3 Parameterization of the Internal Combustion Engine—Exploded View

1.3 Mechanism and Motion Analysis

A mechanism is a mechanical device that transfers motion and/or force from a source to an output. It can be an abstraction (simplified model) of a mechanical system represented as a linkage. A linkage consists of links (or bodies), which are connected by connections (or joints), such as a pin joint, to form open or closed chains (or loops; see Figure 1-4). Such kinematic chains, with at least one link fixed, become mechanisms. In this book, all links are assumed rigid. In general, a mechanism can be represented by its corresponding schematic drawing. For example, a slider-crank mechanism represents the engine motion, as shown in Figure 1-5, which is a closed loop mechanism.

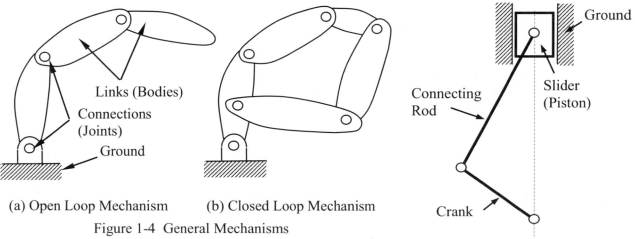

(a) Open Loop Mechanism   (b) Closed Loop Mechanism

Figure 1-4 General Mechanisms

Figure 1-5 Schematic View of the
Engine Motion Model

In general, there are two types of motion problems (among others) that you will need to solve in order to answer general questions regarding mechanism analysis and design. They are kinematic and dynamic problems.

Kinematics is the study of motion without regard for the forces that cause the motion. A kinematic mechanism must be driven by a servo motor (or driver) so that the position, velocity, and acceleration of each link in the mechanism can be analyzed at any given time. Typically, a kinematic analysis must be conducted before dynamic behavior of the mechanism can be simulated properly.

Dynamics is the study of motion in response to externally applied loads. The dynamic behavior of a mechanism is governed by Newton's laws of motion. The simplest dynamic problem is the particle dynamics covered in Sophomore Dynamics—for example, a spring-mass-damper system shown in Figure 1-6. In this case, motion of the mass is governed by the following equation derived from Newton's second law,

$$\sum F = p(t) - kx - c\dot{x} = m\ddot{x} \tag{1.1}$$

where (˙) appearing on top of the physical quantity represents time derivative of the quantity, m is the total mass of the block, k is the spring constant, and c is the damping coefficient.

For a rigid body, mass properties (such as the total mass, center of mass, mass moment of inertia, etc.) are taken into account for dynamic analysis. For example, motion of a pendulum shown in Figure 1-7 is governed by the following equation of motion,

$$\sum M = -mg\ell \sin\theta = J\ddot{\theta} = m\ell^2\ddot{\theta} \tag{1.2}$$

where M is the external moment (or torque), J is the mass moment of inertia of the pendulum, m is the pendulum mass, g is the gravitational acceleration, and $\ddot{\theta}$ is the angular acceleration of the pendulum.

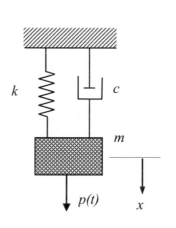
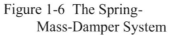

Figure 1-6 The Spring-Mass-Damper System

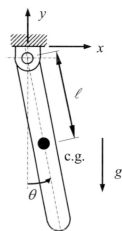

Figure 1-7 A Simple Pendulum

Dynamics of a multi-rigid body system, such as those illustrated in Figure 1-4, is a lot more complicated than the single body problems. Usually, a system of differential and algebraic equations governs the motion and the dynamic behavior of the system. Newton's law must be obeyed by individual bodies in the system at all times. The motion of the system will be determined by the loads acting on the bodies or joint axes (e.g., a torque driving the system). Reaction forces and/or torques at the joint connections hold the bodies together.

Note that in *Mechanism*, you may create a kinematic analysis model, for example, using a servo motor to drive the mechanism before carrying out a dynamic analysis. In this case, position, velocity, and acceleration results may be similar to those of dynamic analysis; however, the inertia of the bodies will be taken into account for dynamic analysis; therefore, reaction forces among other physical quantities will be calculated between bodies.

1.4 *Mechanism* Capabilities

Overall Process

The overall process of using *Mechanism* for designing or analyzing a mechanism consists of three main steps: model creation, analysis, and result visualization, as illustrated in Figure 1-8. Key entities that constitute a motion model include a ground body that is always fixed, bodies that are movable, connections (or joints) that connect bodies, initial conditions that define the initial configuration of the mechanism, servo motors (or drivers) that drive the mechanism for kinematic analysis, and forces and torque that move the components of the mechanism. Most importantly, assembly placement constraints (or assembly mates) must be properly defined for the mechanism so that the motion model captures essential characteristics and closely resembles the behavior of the physical mechanism. More details about these entities are discussed later in this lesson.

The analysis capabilities in *Mechanism* include position (initial assembly), static (equilibrium configuration), kinematic, dynamic, and force balance (to retain the system in a prescribed configuration). For example, a static analysis brings bodies to an equilibrium configuration of the mechanism. More about the analysis capabilities in *Mechanism* will be discussed later in this lesson.

The analysis results can be visualized in various forms. You may animate motion of the mechanism, or generate graphs for more specific information, such as the reaction force of a joint in time domain. You may also query results at specific locations for a prescribed time frame. Furthermore, you may ask for a report on results that you specified, such as the acceleration of a moving body in the time domain.

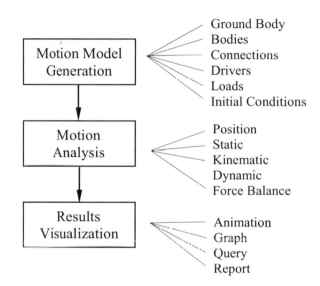

Figure 1-8 General Process of Using *Mechanism*

Operation Mode

Mechanism is embedded into *PTC Creo* (or *PTC Creo Parametric*). It is indeed an integrated module of *Creo*, and transition from *Creo* to *Mechanism* is seamless. All the solid models, assembly constraints, etc. defined in *Creo* are automatically carried over into *Mechanism*. *Mechanism* can be accessed through menus and windows inside *Creo*. The same assembly is used in both *Creo* and *Mechanism*.

Body geometry is essential for mass property computations in motion analysis. In *Mechanism,* all mass properties are ready for use. In addition, the detailed part geometry needed for interference checking is readily available.

User Interface

User interface of the *Mechanism* is identical to that of *Creo*, as shown in Figure 1-9. *PTC Creo* users should find it is straightforward to maneuver in *Mechanism*.

The *Creo* window consists of the following essential elements: command ribbon, *File* pull-down menu, toolbars, navigation area, *Graphics* window, shortcut menus, browser, and status bar.

The ribbon contains the command buttons organized within a set of tabs. On each tab, the related buttons are grouped. Each button on the ribbon consists of an icon and a label. For example, under the *Mechanism* tab, *Mechanism Analysis* ⊠, *Playback* ◀▶, and *Measures* ⊠ buttons are grouped under *Analysis*. The major command buttons in *Mechanism* and their functions are summarized in Table 1-1.

The *File* (pull-down) menu located at the upper-left corner of the *Creo* window supports functions, such as *Open*, *Save*, and *Close*, for managing files and models, and preparing models for distribution.

The *Quick Access* toolbar is available regardless of which tab is selected on the ribbon. By default, it is located at the top of the *Creo* window. It provides quick access to frequently used buttons, such as buttons for opening and saving files, undo, redo, regenerate, close windows, switch windows, and so on. In addition, you can customize the *Quick Access* toolbar to include other frequently used buttons and cascading lists from the ribbon.

The *Graphics* toolbar is embedded at the top of the *Graphics* window. The buttons on the *Graphics* toolbar control the display of model. You can hide or display the buttons on the toolbar. You can change the location of the toolbar by right-clicking and choosing a location from the shortcut menu.

The shortcut menu is contextual user interface relative to the selected object. Press the right mouse button until you see the shortcut menu, for example, the servo motor shown in Figure 1-9.

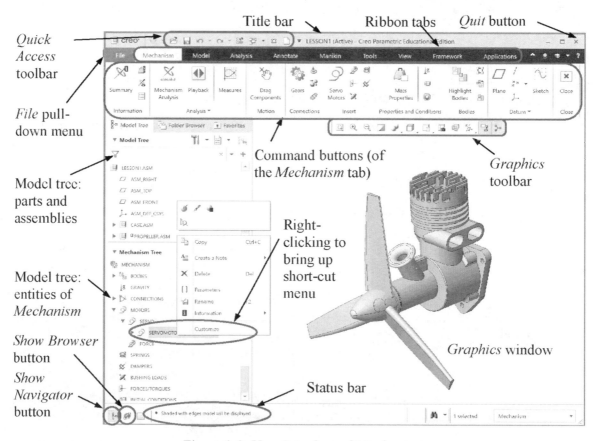

Figure 1-9 User Interface of *Mechanism*

The navigator on the left of the *Graphics* window includes the *Model Tree*, *Layer Tree*, *Detail Tree*, *Folder* browser, and *Favorites*. The *Model Tree* tab [icon] on the status bar (located at the lower left corner of the *Creo* window) controls the display of the navigator. When the *Mechanism* button [icon] is clicked under the *Applications* tab, the model tree is split into two; entities of parts and assemblies are listed in the upper tree, and entities of *Mechanism* are listed in the lower tree, as shown in Figure 1-9.

The *Graphics* window, to the right of the navigator, displays the motion model with which you are working.

The *Creo* browser provides access to internal and external Web sites. Clicking the *Show Browser* button [icon] next to the status bar controls the display of the browser.

Table 1-1 The Major Buttons and Groups of *Mechanism*

Group	Symbol	Name	Function
Analysis	[icon]	Mechanism Analysis	Define and run an analysis.
	[icon]	Playback	Play back the results of your analysis run. You can also save or export the results or restore previously saved results.
	[icon]	Measures	Create measures, and select measures and result sets to display. You can also graph the results or save them to a table.
Motion	[icon]	Drag Components	Drag components in an assembly.
Connections	[icon]	Gears	Create gear pairs.
	[icon]	Cams	Create a cam-follower connection.
	[icon]	3D Contacts	Define a 3D contact between two parts in different bodies.
	[icon]	Belts	Create a belt and pulley system.
Inserts	[icon]	Servo Motors	Define a servo motor (driver).
	[icon]	Force Motors	Define a new force motor.
	[icon]	Forces/Torques	Define a force or a torque.
	[icon]	Bushing Loads	Create a Bushing Load feature to simulate movement between two bodies.
	[icon]	Springs	Define a new spring.
	[icon]	Dampers	Define a new damper.
Properties and Conditions	[icon]	Mass Properties	Specify mass properties for a part or specify density for an assembly.
	[icon]	Gravity	Define gravity.
	[icon]	Initial Conditions	Specify initial position snapshots, and define the velocity initial conditions for a point, motion axis or body.
	[icon]	Termination Conditions	Specify conditions to terminate the motion analysis.

The status bar is located at the bottom of the *Creo* window. When applicable, the status bar displays the controls, such as the display of the navigation area (by clicking the *Show Navigator* button 🔲), and one-line messages related to work in a window. You may right-click in the message area and then click *Message Log* to review past messages.

Defining Motion (or Motion Simulation) Entities

The basic entities of a motion (also called simulation in this book) model created in *Mechanism* consist of ground, bodies, connections, initial conditions, drivers, and loads. Each of the basic entities will be briefly introduced. More details can be found in later lessons.

Ground Body

A ground (or ground body) represents a fixed entity in space. The root assembly is always fixed; therefore, it becomes ground body by default (or part of the ground body). Also, the datum coordinate system of the root assembly is assigned as the *WCS* (World Coordinate System). All datum features and parts fixed to the root assembly are part of the ground body.

Bodies

A body represents a single rigid component (or link) that moves relative to the other body (or bodies in some cases). A body may consist of several *Creo* parts fully constrained using placement constraints. A body contains a local coordinate system (*LCS*), body points (created as datum points), and mass properties. Note that body points are created for defining connections, force applications, etc.

A spatial body consists of three translational and three rotational degrees of freedom (dofs). That is, a rigid body can translate and rotate along the X-, Y-, and Z-axes of a coordinate system. Rotation of a rigid body is measured by referring the orientation of its *LCS* to *WCS*, which is fixed to the ground body.

In *Mechanism*, the *LCS* is assigned automatically, usually, to the default datum coordinate system of the body (either part or assembly), and the mass properties are calculated using part geometry and material properties referring to the *LCS*. Datum axes and points are essential in creating a motion model since they are employed for defining connections and the location of external load application.

Connections

A connection in *Mechanism* can be a joint, cam, or gear that connects two bodies. Typical joints include pin, slider, bearing, ball, cylinder, etc. A connection constrains the relative motion between bodies. Each independent movement permitted by a connection is called a (free) degree of freedom (dof). The degrees of freedom that a connection allows can be translation and rotation along three perpendicular axes, as shown in Figure 1-10.

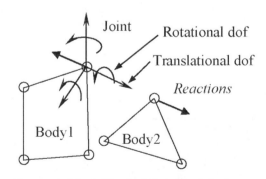

Figure 1-10 A Typical Joint in *Mechanism*

There are two types of options available for assembling parts: *User Defined* (joint) and *Automatic* (placement). The *User Defined* type consists of commonly employed joints for defining motion models, such as rigid, pin, ball, bearing, planar, cylinder joints, etc., which directly correspond to connections implemented in the physical mechanism. The *Automatic* constraints are placement constraints we employed for assembling parts, such as coincident, concentric, etc. Both types eliminate prescribed degrees of freedom between components. For example, a ball joint may be created simply by coinciding two datum points in their respective bodies, allowing all three rotational dofs. Some of the joints are directly equivalent to placement constraint. For example, mating (or aligning, coinciding) two planar faces is equivalent to defining a planar joint. However, instead of completely fixing all the movements, certain dofs (translations and/or rotations) are left to allow the desired movement for a motion model.

The connections produce equal and opposite reactions (forces and/or torques) on the bodies connected. The symbol of a given joint tells the translational and/or rotational dof that the joint allows in regard to movement. Understanding the four basic joint symbols shown in Figure 1-11 will enable you to read existing joints in motion models. More details about joint types available in *Mechanism* will be discussed in later lessons. A complete list of joints available in *Mechanism* can be found in Appendix A.

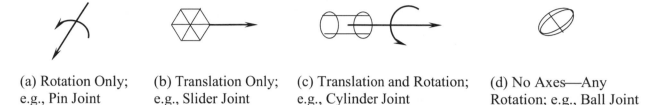

(a) Rotation Only;　　　(b) Translation Only;　　(c) Translation and Rotation;　　(d) No Axes—Any
e.g., Pin Joint　　　　　e.g., Slider Joint　　　　e.g., Cylinder Joint　　　　　Rotation; e.g., Ball Joint

Figure 1-11　The Four Basic Joint Symbols

Degrees of Freedom

As mentioned earlier, an unconstrained body in space has six degrees of freedom: three translational and three rotational. When joints are added to connect bodies, constraints are imposed to restrict the relative motion between them. For example, a pin joint allows one rotational motion between bodies. As defined in the engine example shown in Figure 1-12, joint *Pin1* restricts movement on five dofs so that only one rotational motion is allowed between the propeller assembly and the ground body (*case.asm*).

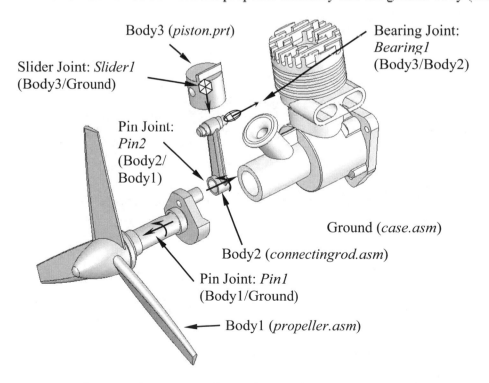

Figure 1-12　An Example Motion Model in Exploded View

For a given motion model, you can determine its number of degrees of freedom using the Gruebler's count. *Mechanism* uses the following equation to calculate the Gruebler's count:

$$D = 6M - N - O \qquad\qquad (1.3)$$

where D is the Gruebler's count representing the overall free degrees of freedom of the mechanism, M is the number of bodies excluding the ground body, N is the number of dofs restricted by all joints, and O is the number of motion drivers (or servo motors) defined in the system.

The single-piston engine shown in Figure 1-12 consists of three bodies (excluding the ground body), two pin joints, one slider joint, and one bearing joint. A pin or slider joint removes five degrees of freedom, and a bearing joint removes two dofs. In addition, a servo motor is added to the rotational dof of the joint *Pin1*. Therefore, according to Eq. 1.3, the Gruebler's count for the engine example is

$$D = 6{\times}3 - (3{\times}5 + 1{\times}2) - 1{\times}1 = 0$$

In general, a valid motion model should have a Gruebler's count of *0*. However, in creating motion models, some joints remove redundant dofs. For example, two hinges, modeled using two pin joints, support a door. The second pin joint adds five redundant dofs. The Gruebler's count becomes

$$D = 6{\times}1 - 2{\times}5 = -4$$

For kinematic analysis, the Gruebler's count must be equal to or less than *0*. The solver recognizes and deactivates redundant constraints during analysis. For a kinematic analysis, if you create a model and try to animate it with a Gruebler's count greater than *0*, the animation will not run and an error message will appear.

If the Gruebler's count is less than zero, the solver will automatically remove redundancies. In this engine example, if the bearing joint between the connecting rod and the piston pin is replaced by a pin joint, the Gruebler's count becomes

$$D = 6{\times}3 - 4{\times}5 - 1{\times}1 = -3$$

To get the Gruebler's count to zero, it is often possible to replace joints that remove a large number of constraints with joints that remove a smaller number of constraints and still restrict the mechanism motion in the same way. *Mechanism* detects the redundancies and ignores redundant dofs in all analyses. In dynamic analysis, the redundancies lead to an outcome with a possibility of incorrect reaction results, yet the motion is correct. For complete and accurate reaction forces, it is critical that you eliminate redundancies from your mechanism. The challenge is to find the joints that will impose non-redundant constraints and still allow for the intended motion. When this is not feasible, cautions must be exercised in checking results of dynamic simulation, especially for reaction forces. Examples included in this book should give you some ideas in choosing proper joints.

Loads

Loads are used to drive a mechanism. Physically, loads are produced by motors, springs, dampers, gravity, etc. A load entity in *Mechanism* can be a force or torque. The force and torque are represented by an arrow and double-arrow symbols, respectively, as shown in Figures 1-13 and 1-14. Note that a load can be applied to a body, a point in a body, or between two points in different bodies.

Figure 1-13 The Force Symbol   Figure 1-14 The Torque Symbol

Drivers or Servo Motors

Drivers or servo motors are used to impose an intended motion on a mechanism. Servo motors cause a specific type of motion to occur between two bodies in a single degree of freedom. Servo motors specify position, velocity, or acceleration as function of time; and can control either translational or rotational motion. The driver symbol is shown in Figure 1-15.

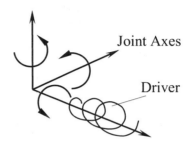

Note that a driver must be defined along a movable axis of the joint you select. Otherwise, no motion will occur. When properly defined, drivers will account for the remaining dofs of the mechanism calculated using Eq. 1.3.

An example of a motion model created using *Mechanism* is shown in Figure 1-12. In this engine example, twenty-six *Creo* parts are grouped into four bodies. In addition, four joints plus a driver are defined for a kinematic analysis. You may open this model and animate its motion by following the steps described in Section 1.5.

Figure 1-15 The Driver
(Servo motor) Symbol

Types of Mechanism Analyses

There are five analysis options supported in *Mechanism*: position, static, kinematic, dynamic, and force balance.

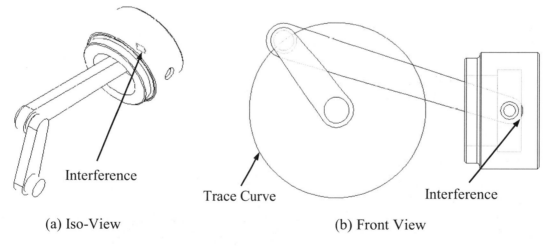

(a) Iso-View   (b) Front View

Figure 1-16 Trace Curve and Interference Checking of a Position (or Assembly)
Analysis for a Slider-Crank Mechanism

A position analysis is a series of assembly analyses driven by servo motors. Only motion axes or geometric servo motors can be included in position analyses. Force motors do not appear in the list of possible motor selections when adding a motor for a position analysis. A position analysis simulates the mechanism's motion, satisfying the requirements of your servo motor profiles and any joint, cam-follower, slot-follower, or gear-pair connections, and records position data for the mechanism's various components. It does not take force and mass into account when performing the analysis. Therefore, you do not have to specify mass properties for your mechanism. Dynamic entities in the model, such as springs, dampers, gravity, forces/torques, and force motors, do not affect a position analysis. You may use a position analysis to study positions of components over time, interference between components, and trace curves of the mechanism's motion (for example, see Figure 1-16).

Static analysis is used to find the rest position (equilibrium condition) of a mechanism, in which none of the bodies are moving. Static analysis is related to mechanical advantage—for example, how much load can be resisted by a driving motor. A simple example of the static analysis is shown in Figure 1-17.

Motion analysis consists of mainly kinematic and dynamic analyses. As discussed earlier, kinematics is the study of motion without regard for the forces that cause the motion. A mechanism can be driven by a servo motor for a kinematic analysis, where the position, velocity, and acceleration of each link of the mechanism can be analyzed at any given time. For example, a servo motor drives a mechanism shown in Figure 1-18 at a constant angular velocity ω.

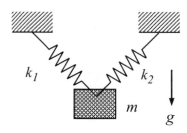

Figure 1-17 Static Analysis

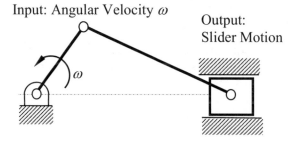

Figure 1-18 Kinematic Analysis

Dynamic analysis is used to study the mechanism motion in response to loads, as illustrated in Figure 1-19. This is the most complicated and common but usually more time-consuming analysis.

Force balance calculates the required force and torque to retain the system in a prescribed configuration.

Viewing Results

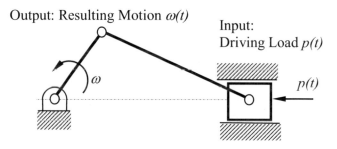

Figure 1-19 Dynamic Analysis

In *Mechanism*, results of the motion analysis can be realized using animations, graphs, reports, and queries. Animations show the configuration of the mechanism in consecutive time frames. Animations will give you a global view on how the mechanism behaves; for example, a motion animation of a single-piston engine shown in Figure 1-20.

You may choose a joint or a datum point in the motion model to generate result graphs; for example, the graph in Figure 1-21 shows the angular position of a simple pendulum example (please see *Lesson 4* for more details). These graphs give you a quantitative understanding of the behavior of the mechanism. You may also pick a data point on the graph to query the results of your interest at a specific time frame. In addition, you may ask *Mechanism* for a report that includes a complete set of results output in the form of numerical data.

In addition to the capabilities discussed above, *Mechanism* allows you to check interference between bodies during motion (more to be discussed in *Lesson 5*). Furthermore, the reaction forces calculated can be used to support structural analysis using, for example, *Creo Simulate*, a p-version finite element analysis module of *PTC Creo*.

1.5 Open Lesson 1 Model

A motion model for the single piston engine model shown in Figure 1-1 has been created for you. Download the files from www.sdcpublications.com, unzip them, and locate the engine assembly file (*lesson1.asm*) under *Lesson 1* folder. Copy *Lesson 1* folder to your computer.

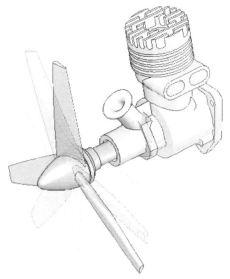

Figure 1-20 Motion Animation

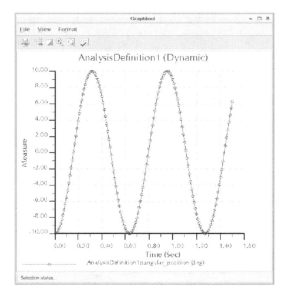

Figure 1-21 A Sample Result Graph

Start *PTC Creo*, select *Lesson 1* folder as the working directory by choosing *File > Manage Session > Select Working Directory*, and open the assembly model: *lesson1.asm*. You should see an assembled engine model similar to that of Figure 1-1.

To enter *Mechanism*, simply click the *Applications* tab on top of the *Graphics* window, and click the *Mechanism* button .

You should see *Mechanism* window layout similar to that of Figure 1-9. To replay results click the *Playback* button on top. The *Playbacks* dialog box (Figure 1-22) appears. In the *Playbacks* dialog box, click the *Open* button , and select the previously saved playback file *AnalysisDefinition1.pbk* (this file is included in the *Lesson 1* folder). Click the *Play current result set* button at the top left corner. The *Animate* dialog box (Figure 1-23) opens. Click the *Play* button to play the motion of the engine. You should see the motion animation similar to that of Figure 1-20.

To close the model, choose *File > Close*. From the *Home* tab, click the *Erase Not Displayed* button in the *Data* group. The *Erase Not Displayed* dialog box opens (Figure 1-24). Click *OK* to erase all parts and assemblies temporarily stored in the memory.

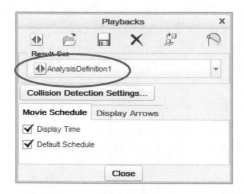

Figure 1-22 The *Playbacks* Dialog Box

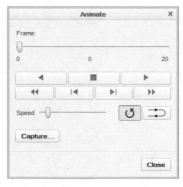

Figure 1-23 The *Animate* Dialog Box

Figure 1-24 The *Erase Not Displayed* Dialog Box

1.6 Motion Examples

Numerous motion examples will be introduced in this book to illustrate step-by-step details of modeling, analysis, and result visualization capabilities in *Mechanism*. We will start with a simple ball throwing example. This example will give you a quick start and a quick run-through on *Mechanism*. *Lessons 3* through *8* focus on modeling and analysis of basic mechanisms. In these lessons, you will learn various joint types, including pin, slider, rigid, etc.; connections, including springs, gears, cam-followers; drivers and forces; various analysis types; and measures and results. *Lessons 9* and *10* are application lessons, in which real-world mechanisms will be introduced to show you how to apply what you learn to more complicated applications. All examples and main topics to be discussed in each lesson are summarized in the Table 1-2.

Table 1-2 Summary of Lessons and Motion Examples in this Book

Lesson	Title	Example	Problem Type	Things to Learn
1	Single-Piston Engine		Multibody Kinematic Analysis	1. General introduction to *Creo Mechanism*
2	Ball Throwing Example		Particle Dynamics	1. This lesson offers a quick run-through on modeling and analysis capabilities in *Mechanism*. 2. You will learn the process of using *Mechanism* to construct a motion model, run analysis, and visualize the motion analysis results. 3. Simulation results are verified using analytical equations of motion.

Table 1-2 Summary of Lessons and Motion Examples in this Book (Cont'd)

Lesson	Title	Example	Problem Type	Things to Learn
3	Spring-Mass System		Particle Dynamics	1. This is a classical spring-mass system example you learned in *Dynamics*. 2. You will learn how to create a mechanical spring, align the block with the slope surface, and add an external force to pull the block. 3. Simulation results are verified using analytical equations of motion.
4	A Simple Pendulum		Particle Dynamics	1. This lesson goes more in-depth about creating joints in *Mechanism*. Pin and rigid joints will be introduced. 2. Simulation results are verified using analytical equations of motion.
5	A Slider Crank Mechanism		Multibody Kinematic and Dynamic Analyses	1. This lesson uses a slider-crank mechanism to introduce more joint types as well as conduct position, kinematic, and dynamic analyses. 2. In addition to joints, you will learn to create drivers for a kinematic analysis. 3. The interference checking capability will be discussed. 4. Kinematic analysis results are verified using analytical equations of motion.
6	A Compound Spur Gear Train		Gear Train Analysis	1. This lesson focuses on simulating motion of a spur gear train system. 2. You will learn how to use *Mechanism* to create gear connections, analyze the gear train, and define measures for gears. 3. Simulation results are verified using analytical equations.
7	Planetary Gear Train Systems		Planetary Gear Train Analysis	1. This lesson is similar to *Lesson 6* but focuses on planetary gear trains. 2. Both single- and multi-gear systems will be discussed. 3. Some simulation results are found incorrect compared with those obtained from analytical equations.
8	Cam and Follower		Multibody Dynamic Analysis	1. This lesson discusses cam and follower joint. 2. An inlet or outlet valve system of an internal combustion engine will be created and simulated. 3. Kinematic and dynamic characteristics of the valve will be monitored in the motion simulation of the system. 4. Design of the system will be discussed.

Table 1-2 Summary of Lessons and Motion Examples in this Book (Cont'd)

Lesson	Title	Example	Problem Type	Things to Learn
9	Assistive Device for Wheelchair Soccer Game		Multibody Dynamic Analysis	1. This is an application lesson. This lesson shows you how to assemble and simulate motion of an assistive device for playing wheelchair soccer game. 2. Numerous joints, spring, and force will be created for the system. 3. Measures will be defined to assess the design of the system.
10	Kinematic Analysis for Racecar Suspension		Multibody Kinematic Analysis	1. This is the second and the last application lesson of the book. A quarter of a racecar suspension will be employed for kinematic analyses. 2. A road profile will be modeled by using a cam with prescribed profile. The cam will be connected to the tire using a cam-follower connection. 3. Various measures, including the camber angle, will be introduced to assess the design of the suspension system.

Note that example files have been prepared for you to go through all the lessons. In addition to _PTC Creo_ parts and assemblies, each lesson folder contains complete motion models as well as simulation result files. You may want to open the motion models and review the simulation results; e.g., play motion animations, to become more familiar with the simulations before going through the lessons.

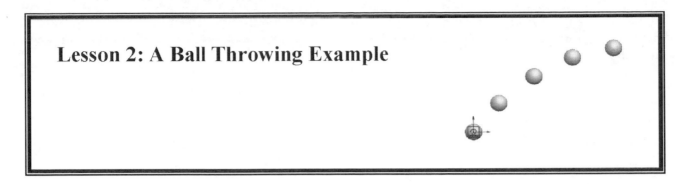

2.1 Overview of the Lesson

The purpose of this lesson is to provide readers a quick start on using *Creo Mechanism*. This example simulates a ball thrown with an initial velocity. Due to gravity, the ball will travel following a parabolic trajectory, as depicted in Figure 2-1. In this lesson, you will learn how to create a motion model to simulate the ball motion, run a dynamic analysis, and animate the ball motion. Dynamic analysis results of this example can be verified using particle dynamics theory you learned in Physics. We will review the equation of motion, calculate the position and velocity of the ball, and compare our calculations with results obtained from *Mechanism*. Validating results obtained from computer simulations is extremely important. *Mechanism* is not foolproof. It requires a certain level of experience and expertise to master the software. Before you arrive at that level, it will be very helpful to verify the simulation results whenever possible. Verifying the simulation results will increase your confidence in using the software and prevent you from being occasionally fooled by the erroneous simulations produced by the software. Note that very often the erroneous results are due to modeling errors.

2.2 The Ball Throwing Example

Physical Model

The physical model of the ball example is simple. The ball is made of *STEEL* with a radius of 0.5 in. Note that the default unit system in *Creo*, i.e., in-lb$_m$-sec, is assumed. The gravitational acceleration is 386 in/sec^2. You may check or change the unit system by choosing from the pull-down menu

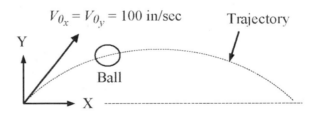

Figure 2-1 The Ball Throwing Example

File > Prepare > Model Properties

Details about changing the unit system can be found in later lessons; for example, *Lesson 4.*

For this lesson, the solid model of the ball has been created for you in *Creo*. You can find the part file at the publisher's web site (www.sdcpublications.com). As mentioned in *Lesson 1*, datum features are extremely important in creating motion models. *Mechanism* converts all datum features of the root assembly to be ground body and requires users to assemble parts (or subassemblies) to the ground body using proper joints (or placement constraints). Joints are similar to the placement constraints, except that joints intentionally allow prescribed movement between bodies, which resembles physical motion of the system. *Mechanism* also converts the assembly datum coordinate system into the World Coordinate System (*WCS*) for the motion model. Note that *WCS* is fixed to the ground body and serves as the ultimate reference for the motion model. Also, joints and forces are usually defined at datum features, such as points and axes. Therefore, we must pay close attention to all datum features created in parts and assemblies.

Creo Part and Assembly

The ball assembly consists of one single part: the ball (see Figure 2-2a). The ball part has a revolved solid feature, a datum axis, three datum planes (*FRONT*, *TOP*, and *RIGHT*), and a datum coordinate system (*PRT_CSYS_DEF*). Also, a datum point *PNT0* is created at the origin of the coordinate system located at the center of the ball. We will create a new assembly and bring the ball part into the assembly.

In the root assembly, you will be given three datum planes (*ASM_RIGHT*, *ASM_TOP*, and *ASM_FRONT*) and an assembly coordinate system (*ASM_DEF_CSYS*). We will create a datum point (*APNT0*) at the origin of the assembly coordinate system (Figure 2-2b) and coincide it (using *coincident* placement constraint) with *PNT0* in the ball to define the initial position of the ball. Note that the ball will be brought into the assembly by creating a planar joint. The planar joint will be defined by coinciding two datum planes: *FRONT* (ball) and *ASM_FRONT* (assembly), which are normal to the z-axis of the respective coordinate systems, *PRT_CSYS_DEF* and *ASM_DEF_CSYS*. As a result, the ball is free to move on the *X-Y* plane.

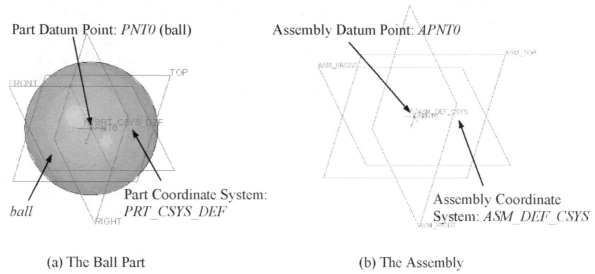

(a) The Ball Part   (b) The Assembly

Figure 2-2 The Ball Part and Assembly

As you are aware, *Creo* dynamically changes the display of the model as you change the view. For example, as you spin, pan, zoom or rotate an object, you see the object changes as you move/drag the mouse. In *Creo*, spin, pan, zoom and rotate are achieved by clicking the middle mouse button (*MB2*) in combination with either the *SHIFT* or *CTRL* key. You may perform the view functions following the instructions shown in Table 2-1. Also, you may place the pointer over a focused area in the *Graphics* window, and then spin the mouse wheel to zoom in/out the object.

Table 2-1 The View Functions Using *MB2*

Spin	*MB2*+Drag
Pan	*MB2*+*SHIFT*+Drag
Zoom	*MB2*+*CTRL*+Drag vertically, or simply scroll forward/backward *MB2*
Rotate	*MB2*+*CTRL*+Drag horizontally

Motion Model

In this example, the ball is the only movable body. A planar joint will be defined between the ball and the ground body (Figure 2-3). The planar joint will restrain the ball to move only on the *X-Y* plane. As mentioned earlier, the ball will be thrown with an initial velocity: in this case, $V_{0_x} = V_{0_y}$ = 100 in/sec. A gravitational acceleration −386 in/sec^2 will be defined in the *Y*-direction of the *WCS* (that is, the assembly coordinate system *ASM_DEF_CSYS*). The ball will reveal a parabolic trajectory due to gravity. The overall simulation will be around 0.52 seconds before the ball hits the ground. We will define measures to graph the *X*- and *Y*-positions of the ball. These measures will be defined at the datum point *PNT0* of the ball. Measures are important for understanding motion analysis results. We discuss steps for defining measures in this lesson. More about measures can be found in Appendix B.

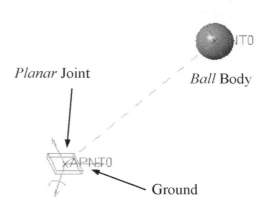

Figure 2-3 The Ball Motion Model

2.3 Using *Mechanism*

Creating an Assembly

Start *Creo*. On the *Home* tab above the *Graphics* window—see Figure 2-4(a)—click *Select Working Directory* button (a command button of the *Data* group) and choose a folder where the part file *ball.prt* is located. Then, click *New* button from the *Data* group to create a new assembly model: *ball_throwing* (or any assembly name you prefer).

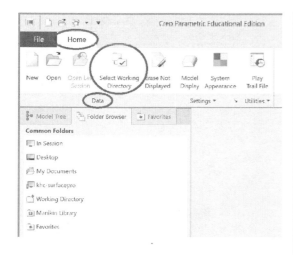

(a) The Options Under the *Home* Tab

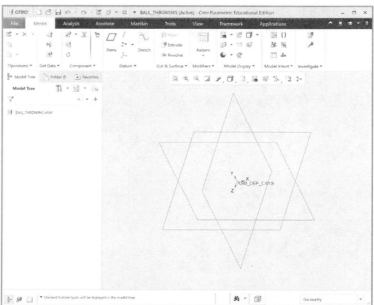

(b) The Default Datum Features in a New Assembly

Figure 2-4 The *Creo* Main Window

You should see three assembly datum planes (*ASM_FRONT*, *ASM_TOP*, and *ASM_FRONT*) and one coordinate system (*ASM_DEF_CSYS*) appear in the *Graphics* window, as shown in Figure 2-4(b).

In the *Model Tree*, there is only one entity *BALL_THROWING.ASM*. To list all features, you may first click the *Settings* button, and then choose *Tree Filters* (Figure 2-5). In the *Model Tree Items* dialog box (Figure 2-6), choose *Features*, and then click *OK*. In the *Model Tree*, you should see all features listed, including datum features (Figure 2-7).

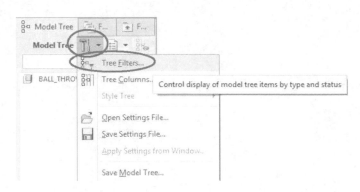

Figure 2-5 The *Tree Filter* option

If you do not see the datum feature labels, for example datum planes, you may click the *View* tab and click *Plane Tag Display* button from the *Show* group to show labels in the *Graphics* window.

We will now create a datum point at the origin of the coordinate system before bringing in the ball part. To create a datum point, you may click the small arrow button next to the *Point* button from the *Datum* group (see Figure 2-8), and choose *Offset Coordinate System* button (2nd button).

We will define the datum point at the origin of the coordinate system *ASM_DEF_CSYS*. In the *Datum Point* dialog box (Figure 2-9a), click the *Reference* field. The field should be highlighted, indicating that it is active and is ready for you to select entities. Pick the coordinate system *ASM_DEF_CSYS* in the *Graphics* window. Then, move the pointer to the *Datum Point* dialog box (Figure 2-9b), and click a table cell right below the table title bar in the top row. You should see that the datum point *APNT0* is listed with a zero offset in all three axes (Figure 2-9b). Click *OK* to accept the datum point definition. Datum point *APNT0* should appear in the *Graphics* window at the origin of the coordinate system.

If you do not see the datum point in the *Graphics* window, you may click *Datum Display Filters* button on the *Graphics* toolbar, and make sure that *Point Display* check box is selected.

Figure 2-6 The *Model Tree Items* Dialog Box

Figure 2-7 Default Assembly Datum Features Listed in *Model Tree*

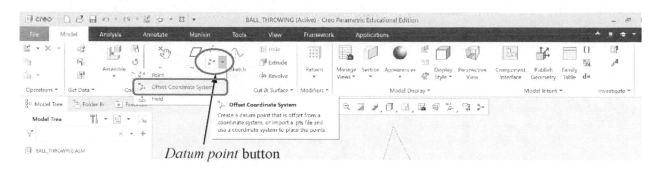

Datum point button

Figure 2-8 Select *Offset Coordinate System* to Create a Datum Point

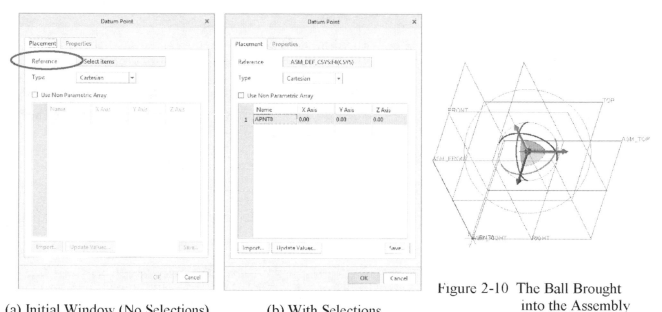

(a) Initial Window (No Selections)   (b) With Selections

Figure 2-9 The *Datum Point* Dialog Box

Figure 2-10 The Ball Brought
into the Assembly

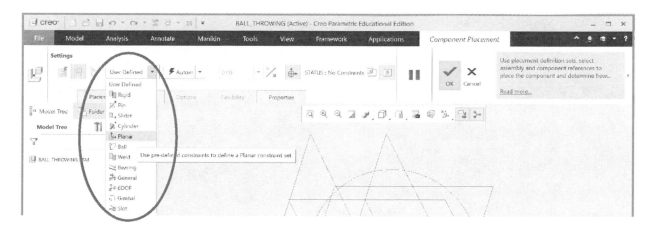

Figure 2-11 Choose *Planar* Joint from the *Component Placement* Dashboard

In case that you do not see the datum point label, you may click the *View* tab and click *Point Tag Display* button ![icon] from the *Show* group to show labels in the *Graphics* window.

Now, we are ready to bring in the ball part. Choose the *Model* tab, and click the *Assemble* button ![icon] of the *Component* group on top of the *Graphics* window. Select *ball.prt* to open. The ball part will appear in the *Graphics* window similar to that of Figure 2-10. In the *Component Placement* dashboard (upper left of the *Graphics* window), choose *Planar* from the *User Defined* list, as shown in Figure 2-11.

Click datum planes *FRONT* (*ball.prt*) and *ASM_FRONT* (assembly) to define the planar joint (see Figure 2-12a). The planar joint symbol ![icon] appears (Figure 2-12b). In addition, you should see the message at the top of the *Graphics* window: *STATUS: Connection Definition Complete*, indicating that the planar joint has been defined successfully. Click the *OK* ![icon] button to the right of the *Component Placement* dashboard to accept the definition.

As discussed in *Lesson 1*, there are two types of options available for assembling parts in the *Component Placement* dashboard. They are *User Defined* (joint) and *Automatic* (placement). The *User Defined* type consists of commonly employed joints for defining motion models, such as the planar joint we just defined. These options offer joint definitions that are directly corresponding to connections implemented in the physical mechanism, such as rigid, pin, cylinder joints, etc. The *Automatic* constraints (or placement constraints) are regular constraints we employed for assembling parts in solid modeling, such as *coincident*, *tangent*, and so on. Both types eliminate prescribed degrees of freedom between components. Some of the joints are directly equivalent to placement constraints. For example, mating two planar faces is equivalent to defining a planar joint. Some joints require a combination of placement constraints. For example, a pin joint requires an axis align and face mate placement constraints. You may use the *Convert* button ![icon] at the top of the *Graphics* window to convert the placement constraints to joints or vice versa. Choose adequate joints and/or placement constraints to define your motion model.

Note that we would like to position the ball with its center point (*PNT0*) coincident with the assembly datum point (*APNT0*). The assembly datum point *APNT0* is fixed to the ground. In fact, all the assembly datum features belong to the ground body. This will be where the ball is positioned before any motion.

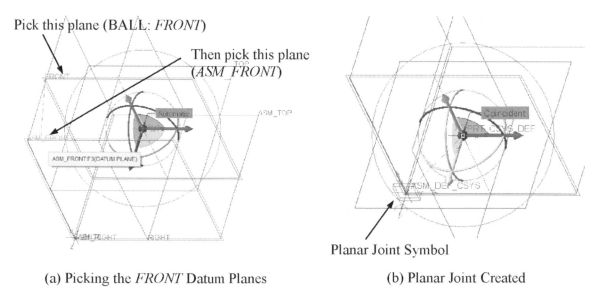

(a) Picking the *FRONT* Datum Planes   (b) Planar Joint Created

Figure 2-12 The Ball Part Being Assembled to the Ground Body

You may move the ball on the *FRONT* plane (normal to the z-axis since a planar joint is defined) by using the *Drag* option in *Creo*. Before dragging the ball, you may want to set the view to be normal to the z-axis. We will choose the *FRONT* view from the saved view list.

To set the view, click the *Saved Orientations* button 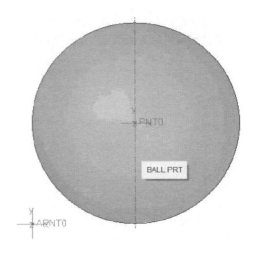 of the *Graphics* toolbar and choose *FRONT*. In the *Graphics* window, you should see a view similar to that of Figure 2-13. Note that datum plane display has been turned off in Figure 2-13.

Figure 2-13 Front View (Datum Plane Display Turned Off)

Click the *Drag Components* button [icon] of the *Component* group at the top of the *Graphics* window. The *Drag* dialog box (Figure 2-14a) appears. Click the ball in the *Graphics* window. Without clicking the mouse again, simply move the mouse to drag the ball on the plane that is normal to the z-axis as expected.

In the *Drag* dialog box, click the expand button (triangular shape next to the label *Snapshots*, see Figure 2-14b) to expand the window. Click the *Constraints* tab and choose the *Align Two Entities* button (first on the left, as shown in Figure 2-14c). Choose two datum points, *PNT0* and *APNT0*. The ball should now be centered at the assembly datum point *APNT0*. Click the *Current Snapshots* button [icon] (see Figure 2-14c) to save a snapshot of the current configuration under the default name: *Snapshot1*. Note that this snapshot defines the initial position of the ball. Before running a motion analysis, make sure you bring this snapshot back in case the ball has been dragged away. Click *Close* to accept the snapshot.

(a) Unexpanded

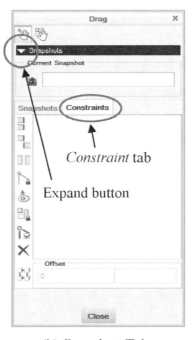

Constraint tab

Expand button

(b) *Snapshots* Tab

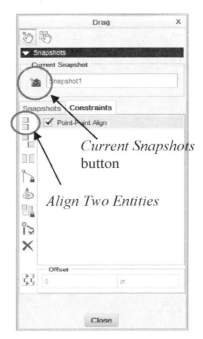

Current Snapshots button

Align Two Entities

(c) *Constraint* Tab

Figure 2-14 The *Drag* Dialog Box

Creating a Dynamic Simulation Model

We are now ready to enter *Mechanism* for creating a dynamic simulation model. To enter *Mechanism*, simply click the *Applications* tab on top of the *Graphics* window, and choose *Mechanism* button.

Note the change of the *Creo* window (see Figure 2-15 with datum planes display turned back on, and display set in the *Standard Orientation*). First, the *Mechanism* function buttons appear on top of the *Graphics* window, providing all the functions for creating motion models, defining and running motion analyses, and visualizing results. In this lesson, we will use gravity, initial conditions, mechanism analysis, playback, and measures buttons. The second change is that the *Model Tree* window is split into two. The added lower half lists motion model entities, including bodies, connections, analysis, etc. It is convenient to review or modify existing entities by clicking the entity and pressing the right mouse button. In addition, a planar joint symbol re-appears at the center of the ball in the *Graphics* window, similar to that of Figure 2-12b.

Next, we define the gravitational acceleration and add an initial velocity $V_{0_x} = V_{0_y} = 100$ in/sec to the ball.

Click the *Gravity* button of the *Properties and Conditions* group (see Figure 2-15) to bring up the *Gravity* dialog box (Figure 2-16). Note that the default value of acceleration (386.088) and default direction (0,–1,0) appear, which are what we need. No change is necessary. Simply click *OK* in the *Gravity* dialog box. You will need to activate the gravity when you define a dynamic analysis (to discuss later).

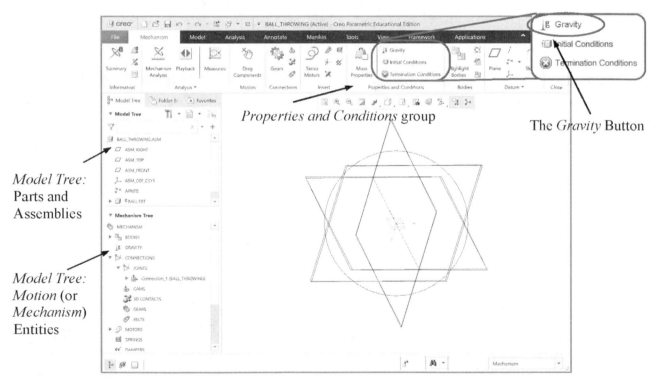

Figure 2-15 Changes in *Mechanism* Window

Click the *Initial Conditions* shortcut button 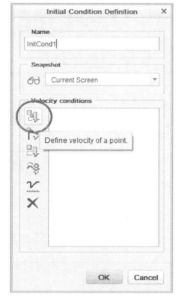 right below the *Gravity* button. In the *Initial Condition Definition* dialog box (Figure 2-17), click the *Define velocity of a point* button (first on the left), and then pick *PNT0* (Note: do not pick *APNT0* since the initial velocity must be defined to the ball). Use the right mouse button to shuffle the overlapped entities (for example, *APNT0* and *PNT0*) in the *Graphics* window and bring up the desired entity to pick. You may also pick the entity from the *Model Tree*. Enter *Magnitude: 141.4* (note that the velocity magnitude is $V = \sqrt{V_{0x}^2 + V_{0y}^2}$), and *1, 1, 0* for *X, Y,* and *Z,* respectively in the *Initial Condition Definition* dialog box (see Figure 2-18). Click *OK* to accept the definition. A direction arrow appears in the *Graphics* window, indicating the direction of the initial velocity, as shown in Figure 2-19.

Creating and Running a Dynamic Analysis

Before running a motion analysis, make sure the ball is brought back to its initial position, as defined in the snapshot (*Snapshot1*) earlier. Click the *Drag Packed Components* button  of the *Motion* group at the top of the *Graphics* window. The *Drag* dialog box should appear with *Snapshot1* listed (Figure 2-20). Double click *Snapshot1* to bring back the *Snapshot1* configuration. The ball should be restored to the position defined in *Snapshot1* (if the ball has been dragged away). Click the *Close* button to close the *Drag* dialog box.

Figure 2-16 The *Gravity* Dialog Box

Figure 2-17 The *Initial Conduction Definition* Dialog Box

Figure 2-18 Initial Condition Defined

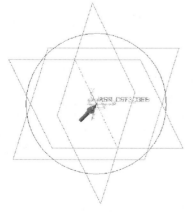

Figure 2-19 The Direction Arrow of the Initial Velocity

Click the *Mechanism Analysis* button [×] of the *Analysis* group. The *Analysis Definition* dialog box (Figure 2-21a) appears. Under *Type*, select *Dynamic*. Leave the default name, *AnalysisDefinition1*. Enter:

Start Time: *0*
End Time: *0.6*
Frame Rate: *100*
Minimum Interval: *0.01*

and click the *I.C. State* radio button (*InitCond1* should be listed).

Note that *Frame Rate* defines number of time frames and the *Minimum Interval* specifies the time interval on which the results will be reported. Note that the *Frame Rate* and *Minimum Interval* are related; i.e., *Frame Rate = 1/Interval*.

Click the *External Loads* tab, click *Enable Gravity* (Figure 2-21b), and then click *Run*.

The progress of the analysis is shown in the *Prompt/Message* window (right below the *Graphics* window), and the ball starts moving following a parabolic trajectory.

Figure 2-20 The *Drag* Dialog Box

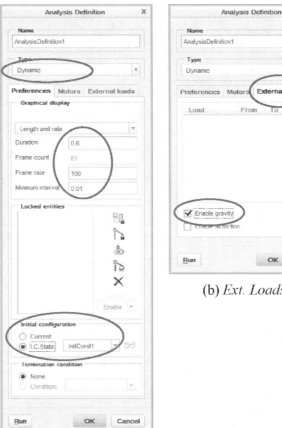

(b) *Ext. Loads* Tab

(a) *Preferences* Tab

Figure 2-21 The *Analysis Definition* Dialog Box

If the ball travels out of the view, you may use the *Refit* button ⬜ on the *Graphics* toolbar or zoom out of the *Graphics* window a few times to locate the ball. Click *OK* to close the *Analysis Definition* dialog box.

Note that the analysis results must be saved as a playback file in order to use them later. We will discuss how to do so next.

Saving and Reviewing Results

To play the motion animation of the simulation results, click the *Playback* button ⬜ of the *Analysis* group on top of the *Graphics* window. The *Playbacks* dialog box (Figure 2-22) opens. In the *Playbacks* dialog box the *AnalysisDefinition1* appears in the *Result Set* field (Figure 2-22). Click the *Play Current Result Set* button ⬜ at the top left corner. The *Animate* dialog box (Figure 2-23) appears. Click the *Play* button ⬜ to play the motion of the ball. Click *Stop* button ⬜ to stop the animation. Again, you may need to refit the screen or zoom out of the *Graphics* window to see the entire trajectory (Figure 2-24). You may want to choose *Shading with Edge* option from the *Display Style* button of the *Graphics* toolbar to see the ball. Click *Close* button to close the *Animate* dialog box.

In the *Playbacks* dialog box (see Figure 2-22), click the *Save* button ⬜ to save your results as a *.pbk* file. In the *Save Analysis Results* dialog box (Figure 2-25), accept the default file name (*AnalysisDefinition1.pbk*) or specify another file name. The default directory is the current working directory. You may also select another directory to save your file. Click *Save* to save the playback file.

You may open an existing *.pbk* file by clicking the *Open* button ⬜ from the *Playbacks* dialog box (Figure 2-22) and selecting the previously saved playback file. Click *Close* to close the *Playbacks* dialog box.

To review results in graphs, click the *Measure* button ⬜ of the *Analysis* group on top of the *Graphics* window. The *Measure Results* dialog box appears (Figure 2-26). In the *Measure Results* dialog box click the *New* button ⬜ (first on the left of the *Measures* table). The *Measure Definition* dialog box appears (Figure 2-27). Enter *X_Position* for *Name*. Under *Type*, select *Position*. Click the *Select* button ⬜ and pick *PNT0* in the *ball* part. Leave *WCS* as the *Coordinate System*. Choose *X-component* for *Component*. Under *Evaluation Method*, leave *Each Time Step*. Click *OK* to accept the definition.

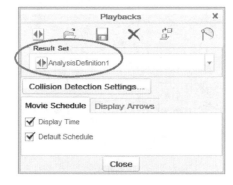

Figure 2-22 The *Playbacks* Dialog Box

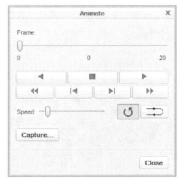

Figure 2-23 The *Animate* Dialog Box

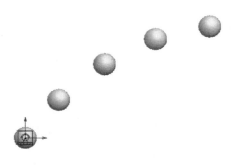

Figure 2-24 Motion Animation

Repeat the same steps to define the *Y_Position* measure. After that, you should see both measures are listed in the *Measure Results* dialog box (Figure 2-28).

Select *AnalysisDefinition1* under *Result Set*. (If you changed the result set name, select the respective name). The *Graph Type* should be *Measure vs. Time* (on top) and the measure values at the current status will appear in the *Measures* table. More about the measures in *Mechanism* can be found in Appendix B.

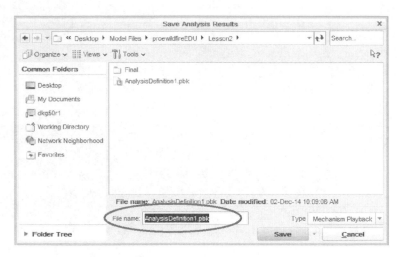

Figure 2-25 The *Save Analysis Results* Dialog Box

Choose both measures by clicking them while pressing the *Shift* key. Click ⬚ on the top left corner to show a graph for the measures. The graph should be similar to that of Figure 2-29.

As shown in Figure 2-29, the *X*-position of the ball is represented in a straight line with a slope of 100 in/sec, which is the *X*-component of the initial velocity. The *Y*-position is a parabolic curve. The ball will hit the ground (when *Y*-position is zero) around 0.52 seconds. This is why we entered 0.6 seconds for the analysis duration. *Mechanism* provides the functionality to terminate the analysis when the ball reaches the ground. We will learn how to do that next. Close the *Graph* window and the *Measure Results* dialog box.

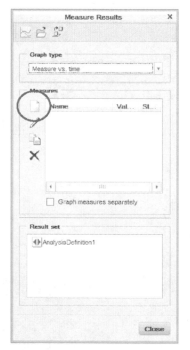

Figure 2-26 The *Measure Results* Dialog Box

Figure 2-27 The *Measure Definition* Dialog Box

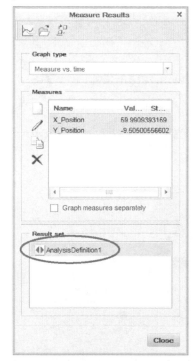

Figure 2-28 The *Measure Results* Dialog Box After

Termination Condition

As seen in Figure 2-29, the *Y*-position of the ball becomes negative after about 0.52 seconds in the simulation. How do we terminate the simulation as soon as the ball hits the ground; i.e., Y-position becomes negative? We may create a termination condition to do so.

Click the *Termination Conditions* button 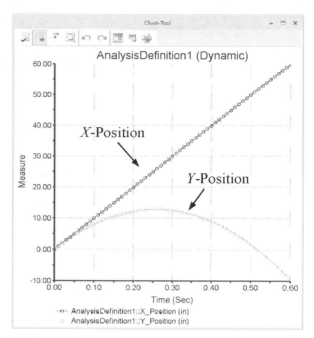 of the *Properties and Conditions* group. In the *Termination Condition Definition* dialog box (Figure 2-30), click the *Insert variable from list* button (Figure 2-30) to open the *Variables* dialog box (Figure 2-31). The two measures, *X_Position* and *Y_Position*, defined previously are listed. Choose *Y_Position* and click *Close*. Back in the *Termination Condition Definition* dialog box (Figure 2-32), *Y_Position* appears. Enter *< 0* next to it and click *OK* (see Figure 2-32).

Figure 2-29 The *X*- and *Y*-Position Graphs

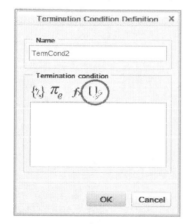

Figure 2-30 The *Termination Condition Definition* Dialog Box

Figure 2-31 The *Variable* Dialog Box

Figure 2-32 The *Termination Condition Definition* Dialog Box with the Condition Defined

In the *Mechanism* model tree, expand *ANALYSIS*, right click *AnalysisDefinition1 (DYNAMICS)*, and click the *Edit Definition* button . In the *Analysis Definition* dialog box, click *Condition* under *Termination condition* (see Figure 2-33). *TermCond1* should appear. Click *OK* to accept the termination condition.

Before re-running the analysis, you may click the *Drag Packed Components* button and double click *Snapshot1* to bring the ball back to its initial configuration.

In the *Mechanism* model tree, right click *AnalysisDefinition1 (DYNAMICS)*, and click the *Run* button . Choose *Yes* to the message: *Selected result set is already available in session. Do you want to overwrite it?* The ball moves and stops at the same elevation as its initial position as expected. You may choose the *FRONT* view from the saved view list to see the animation.

Exporting the Graph

You may export data shown in a graph to a text file or an *Excel* spreadsheet. All you have to do is to click the *Export data to an Excel file* button ⊞ above the graph (for example see Figure 2-29) and enter a filename for the *Excel* file.

Reviewing and Changing the Dynamic Model

From time to time you may need to review or modify a given motion model. Reviewing and changing the motion model is straightforward in *Mechanism*. The best way to do so is choosing options by right clicking motion entities in the *Model Tree*. In the *Model Tree* window, you should see that the entities created for the motion model are listed in the lower half of the window (Figure 2-34).

In order to review or change an entity, simply click the entity name listed in the *Model Tree* window, and use the right mouse button to bring up the options for editing.

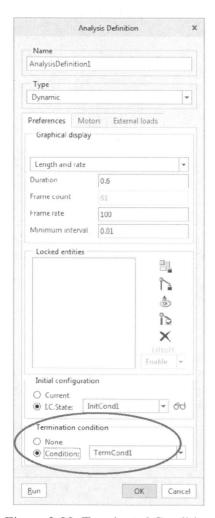

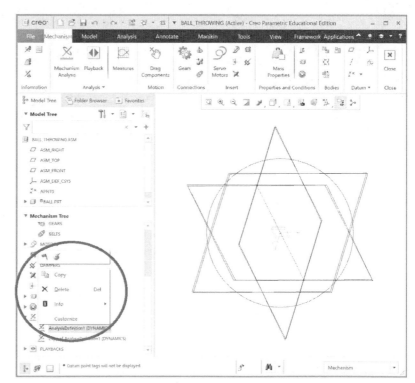

Figure 2-34 The Motion Model Entities Listed in the *Model Tree* Window

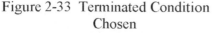

Figure 2-33 Terminated Condition
Chosen

For example, you may bring up the *Analysis Definition* dialog box (Figure 2-21a) by clicking *ANALYSIS* to expand its contents, clicking *AnalysisDefinition1* using the right mouse button, and then choosing the *Edit Definition* button (see Figure 2-34) to bring up dialog box for changes.

Make sure to save your model before exiting from *Creo* by choosing *File* > *Save* from the *File* (pull-down) menu. Then, choose *File* > *Close* to close the model.

Also, make sure you erase parts and assemblies from the memory by clicking from the *Home* tab the *Erase Not Displayed* button in the *Data* group. Click *OK* in the *Erase Not Displayed* dialog box to erase all parts and assemblies temporarily stored in the memory.

2.4 Result Verifications

In this section, we will verify analysis results obtained from *Mechanism* using particle dynamics theory you learned in Sophomore *Physics*.

There are two assumptions that we have to make in order to apply the particle dynamics theory to this ball-throwing problem:

(i) The ball is of a concentrated mass, and
(ii) No air friction is present.

Equation of Motion

It is well-known that the equations that describe the position and velocity of the ball are, respectively,

$$P_x = V_{0_x} t \qquad (2.1a)$$

$$P_y = V_{0_y} t - \frac{1}{2} g t^2 \qquad (2.1b)$$

and

$$V_x = V_{0_x} \qquad (2.2a)$$

$$V_y = V_{0_y} - g t \qquad (2.2b)$$

where P_x and P_y are the *X*- and *Y*-positions of the ball, respectively; V_x and V_y are the *X*- and *Y*-velocities, respectively; V_{0x} and V_{0y} are the initial velocities in the *X*- and *Y*-directions, respectively; and g is the gravitational acceleration.

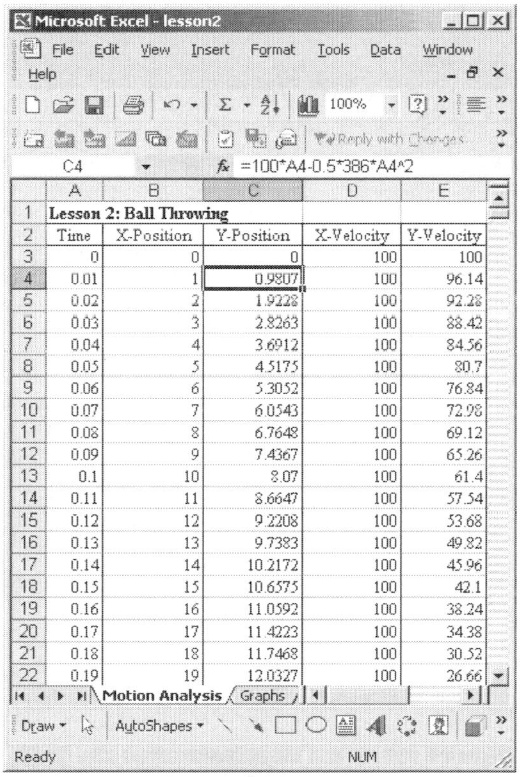

C4 =100*A4-0.5*386*A4^2

	A	B	C	D	E
1	Lesson 2: Ball Throwing				
2	Time	X-Position	Y-Position	X-Velocity	Y-Velocity
3	0	0	0	100	100
4	0.01	1	0.9807	100	96.14
5	0.02	2	1.9228	100	92.28
6	0.03	3	2.8263	100	88.42
7	0.04	4	3.6912	100	84.56
8	0.05	5	4.5175	100	80.7
9	0.06	6	5.3052	100	76.84
10	0.07	7	6.0543	100	72.98
11	0.08	8	6.7648	100	69.12
12	0.09	9	7.4367	100	65.26
13	0.1	10	8.07	100	61.4
14	0.11	11	8.6647	100	57.54
15	0.12	12	9.2208	100	53.68
16	0.13	13	9.7383	100	49.82
17	0.14	14	10.2172	100	45.96
18	0.15	15	10.6575	100	42.1
19	0.16	16	11.0592	100	38.24
20	0.17	17	11.4223	100	34.38
21	0.18	18	11.7468	100	30.52
22	0.19	19	12.0327	100	26.66

Figure 2-35 The *Excel* Spreadsheet

The above equations can be implemented using, for example, *Microsoft Excel*, shown in Figure 2-35, for numerical solutions. As shown in Figure 2-35, Columns B and C show the results of Eqs. 2.1a and b, respectively, with a time interval from 0 to 0.52 seconds and increment of 0.01 seconds. Also, Columns D and E show the results of Eqs. 2.2a and b, respectively. Data in columns B and C are graphed in Figure 2-36. Also, columns D and E are graphed in Figure 2-37. Comparing Figure 2-36 with Figure 2-29, the results obtained from theory and *Mechanism* are very close (they are virtually identical), which means the dynamic model has been created correctly in *Mechanism*, and *Mechanism* does its job and gives us good results. Note that the solution spreadsheet can be found at the publisher's website (filename: *lesson2.xls*).

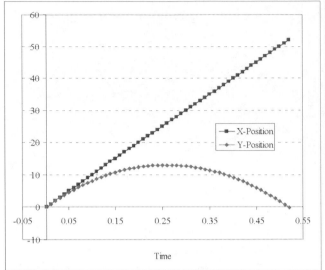

Figure 2-36 Graph of the *X*- and *Y*-Positions of the Ball Obtained from Spreadsheet Calculations

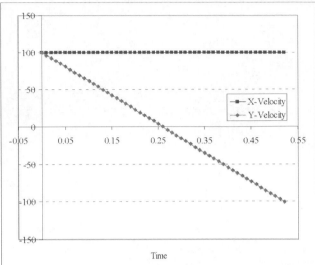

Figure 2-37 Graph of the *X*- and *Y*-Velocities of the Ball Obtained from Spreadsheet Calculations

Exercises:

1. A ball is thrown at an initial velocity of $V_{0_x} = 100$ in/sec from a stand that is 100 in. above the ground, as shown in Figure E2-1. The radius of the ball is 0.1 in., and the material is steel.

    (i)   Create a dynamic simulation model using *Mechanism* to simulate the trajectory of the ball. Report position, velocity, and acceleration of the ball at 0.05 seconds in both vertical and horizontal directions obtained from *Mechanism*.

    (ii) Derive and solve the equations that describe the position and velocity of the ball. Compare your solutions with those obtained from *Mechanism*.

    (iii) Calculate the time for the ball to reach the ground. Compare your calculation with the simulation obtained from *Mechanism* by defining a termination condition.

2. A 1"×1"×1" block slides from top of a $45°$ slope (due to gravity) without friction, as shown in Figure E2-2. The material of the block is *AL2014*.

    (i)   Create a dynamic simulation model using *Mechanism* to analyze motion of the block. Report position, velocity, and acceleration of the block in both vertical and horizontal directions at 0.125 seconds obtained from *Mechanism*.

    Hint: create a planar joint between the block and the slope face.

    (ii) Derive and solve the equation of motion for the system. Compare your solutions with those obtained from *Mechanism*.

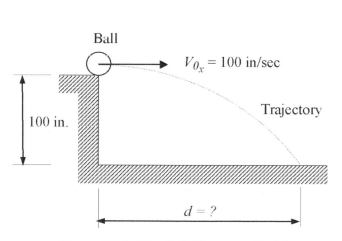

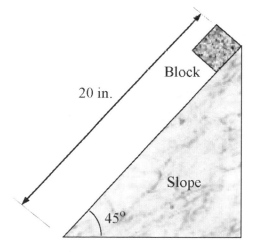

Figure E2-1 The Ball Throwing Problem   Figure E2-2 The Block Sliding Problem

Notes:

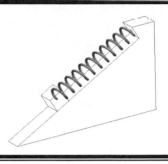

Lesson 3: A Spring-Mass System

3.1 Overview of the Lesson

In this lesson, we will create a simple spring-mass system to simulate its dynamic responses. A schematic of the system is shown in Figure 3-1, in which a steel block of 1in.× 1in.×1in. is sliding along a 30° slope with a spring connecting it to the top end of the slope. The block will slide back and forth along the slope under different scenarios. In this lesson, you will learn to create the spring-mass model, run motion analyses, and visualize the analysis results. The analysis results of the spring-mass example can be verified using particle dynamics theory. Similar to *Lesson 2*, we will formulate the equation of motion, solve the differential equations, graph positions of the block, and compare our calculations with results obtained from *Mechanism*.

3.2 The Spring-Mass System

Physical Model

Note that the default unit system in-lb_m-sec will be used for this example. The spring constant and free length are (in the in-lb_m-sec unit system) $k = 20$ lb_m/sec^2 (which is different from $lb_f/in.$ we are familiar with) and $U = 3$ in., respectively. Please refer to Appendix C for more details regarding unit systems, especially the mass and force units in the English system, i.e., lb_m and lb_f. No friction is assumed between the block and the slope.

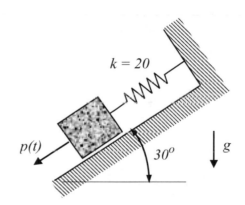

Figure 3-1 The Spring-Mass System

Three scenarios are included in this lesson. The first scenario assumes a free vibration, where the block is stretched 1 in. downward along the 30° slope, where no gravitational acceleration exists. The second scenario is identical to the first one except a gravitational acceleration $g = 386$ in/sec² is assumed. In the third scenario, an external force $p(t) = 10 \cos 2t$ lb_m in/sec² is applied to the block with the gravitational acceleration, as shown in Figure 3-1. Note that the force magnitude is very small (10 lb_m in/sec² = 0.0259 lb_f). All three scenarios will be simulated using *Mechanism*.

Creo Parts

In this lesson, *Creo* parts of the spring-mass example have been created for you. You can find the model files at the publisher's web site (www.sdcpublications.com). As mentioned in *Lesson 1*, datum features are extremely important in creating a successful motion model. In this lesson, we will use both datum points and datum axes to assist in creating the motion model.

The spring-mass system consists of two parts: the block (*block.prt*) and the ground (*ground.prt*), as shown in Figure 3-2. The *block* part has an extruded solid feature, a datum axis (*A_1*), three datum planes (*FRONT*, *TOP*, and *RIGHT*), a coordinate system (*PRT_CSYS_DEF*), and a datum point *PNT0* at the origin of the coordinate system.

Similarly, the ground part has an extruded solid feature, a datum axis (*A_1*), three datum planes (*FRONT*, *TOP*, and *RIGHT*), a coordinate system (*PRT_CSYS_DEF*), and two datum points (*PNT1* and *PNT2*). Note that the datum axes are created to align the block properly to the ground as part of the initial conditions. The datum points *PNT0* of the block and *PNT1* of the ground are used to define the spring. In addition, *PNT0* of the block and *PNT2* (4 in. downward along the slope) of the ground will be aligned to define the initial position of the block before starting the simulation.

We will create a new assembly called *spring_mass* and bring in these two parts. In the new assembly, you will be given three datum planes (*ASM_RIGHT*, *ASM_TOP*, and *ASM_FRONT*) and an assembly coordinate system (*ASM_DEF_CSYS*). The ground part will be brought into the assembly by aligning (using coincident placement constraint) their respective coordinate systems. The block will be brought in by creating a planar joint between the block and the ground. Note that the planar joint will be defined by mating the bottom face of the block with the slope face of the ground, as shown in Figure 3-2. As a result, the block is free to move on the face of the slope. We will then align the two datum axes (axis *A_1* of the block and *A_1* of the ground) in order to properly constrain the motion of the block, as shown in Figure 3-3.

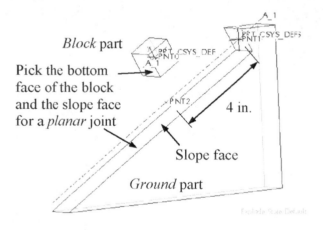

Figure 3-2 The Block and Ground Parts

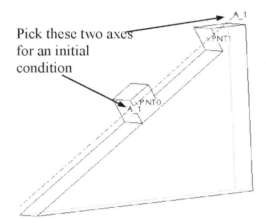

Figure 3-3 Spring Mass Assembly

Motion Model

In this example, the block is the only movable body. A planar joint will be defined between the block and the ground body (*ground.prt*), as shown in Figure 3-4. The planar joint that is co-planar with the slope face will restrain the block to move along the 30° plane. A spring is defined by connecting datum points *PNT0* of the block and *PNT1* of the ground. As mentioned earlier, the spring has a free length $U = 3$ in. The block will be stretched further down to the slope 1 in. from the unstretched configuration. This will be defined by coinciding *PNT0* of the block with *PNT2* of the ground (4 in. from *PNT1* along the slope; see Figure 3-2).

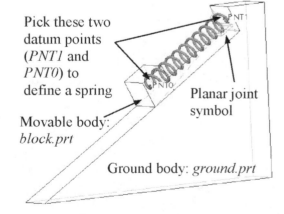

Figure 3-4 Spring Mass Dynamic Model

The spring will be released with zero initial velocity. In the first scenario, no gravity exists. Gravitational acceleration and an external force will be added for the remaining two scenarios.

3.3 Using *Mechanism*

Creating an Assembly

Start *Creo*, select working directory, and create a new assembly: *spring_mass.asm* (or any assembly name you prefer). You should see three assembly datum planes (*ASM_FRONT*, *ASM_TOP*, and *ASM_FRONT*) and one coordinate system (*ASM_DEF_CSYS*).

Click the *Assemble* button of the *Component* group on top of the *Graphics* window, and choose *ground.prt*.

The *ground* part will appear in the *Graphics* window. We will assemble the ground part by coinciding the two datum coordinate systems: *PRT_CSYS_DEF* of the ground part and *ASM_CSYS_DEF* of the assembly.

In the *Component Placement* dashboard (upper left) click the *Placement* button, and choose *Coincident* from the *Constraint Type*, as shown in Figure 3-5. Click the datum coordinate systems, *PRT_CSYS_DEF* (*ground.prt*) and *ASM_DEF_CSYS* (assembly) from the *Graphics* window. Click the *OK* button at right of the *Component Placement* dashboard to accept the definition.

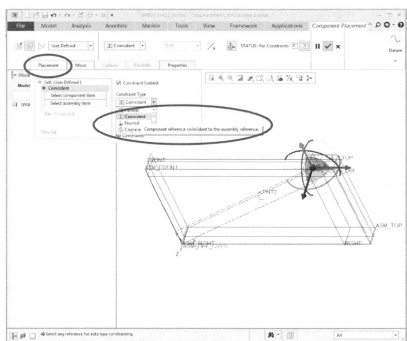

Figure 3-5 Choosing *Coord Sys* from the *Component Placement* Dashboard

Next, we assemble the block to the ground by defining a planar joint.

Click the *Assemble* button and choose the block part. The block will be brought into the *Graphics* window. Note that you may want to turn off some datum feature displays (for example, datum planes) to make the datum entities visible and easier to pick. Move and orient the block to a configuration similar to that of Figure 3-6 by dragging the blue, red, and green arrows and arcs of the *3D dragger* surrounding the part. Most importantly, orient the block so that the two axes, *A_1* of the block and *A_1* of the ground, are roughly aligned (similar to those of Figure 3-6).

Choose the planar joint from the *User Defined* list of the *Component Placement* dashboard. Choose the slope face of the ground and the bottom face of the block (rotate the model to pick the bottom face) for the planar joint. The planar joint symbol will appear. Click the button to the right of the *Component Placement* dashboard to accept the definition.

Note that the block can now be dragged freely on the 30° plane (for example, using the *Drag Components* button), which is not quite yet the motion model we are looking for.

In order to constrain the block motion within the rectangular slope face, we would in general align axes *A_1* in the block and *A_1* in the ground. We will not create such a placement constraint in this case. Instead, we will simply align the two axes using the *Drag Components* capabilities in *Mechanism*. We will learn to do such after creating a spring to connect the block with the ground in the simulation model.

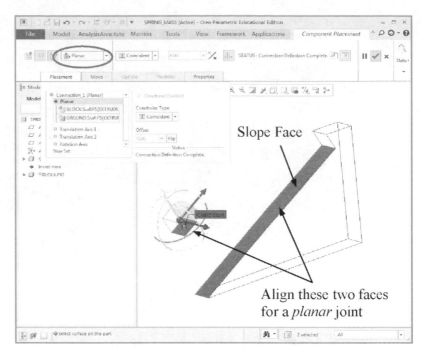

Figure 3-6 Defining a *Planar* Joint

Creating a Dynamic Simulation Model

Now we enter *Mechanism* by clicking the *Applications* tab on top of the *Graphics* window, and choosing *Mechanism* button.

We will first define a spring and set the initial conditions for the motion model.

Click the *Springs* button of the *Insert* group; a new set of selections will appear at the top of the *Graphics* window for defining the spring (see Figure 3-7). Click the *Extension or compression spring* button (the first button from the left—should have been selected by default). Activate the *Select items* field by clicking it. Turn on the datum point display. Then, pick *PNT0* of the block in the *Graphics* window. Drag the handle that appears in the *Graphics* window (see Figure 3-8) and overlap it with *PNT1* (release the mouse button when *PNT1* is highlighted). A spring symbol will appear, connecting *PNT0* (block) and *PNT1* (ground).

Next, enter 20 for spring constant *K* and 3 for the free length *U* from the text fields at the top of the *Graphics* window (see Figure 3-7). Note that the spring constant we enter is very small. The constant $K = 20$ $lb_m/sec^2 = 20/386$ $lb_f/in. = 0.0518$ $lb_f/in.$ It is a very soft spring. We expect to see a relatively low vibration frequency.

Note that you may adjust the spring diameter (only for display purpose) by clicking the *Options* button below the *Select item* field (Figure 3-9). Also, you may change the spring name by clicking the *Properties* button and entering name. Click the button on the right to accept the definition.

In order to properly align the block with the slope face of the ground, we will use the *Drag* command to align the two *A_1* axes, as shown in Figure 3-10. We will also align two datum points, *PNT0* of the block and *PNT2* of the ground, to set up the initial position for the block.

Click the *Drag Components* button 🖐 at the top of the *Graphics* window, and the *Drag* dialog box appears (Figure 3-11, similar to what we learned in *Lesson 2*). Expand the dialog box, click the *Constraints* tab, and choose the *Align Two Entities* button (first on the left). Choose the two datum axes, *A_1* (block) and *A_1* (ground) (see Figure 3-10). The block should now be sitting on top of the slope face and the spring should align properly with the slope face.

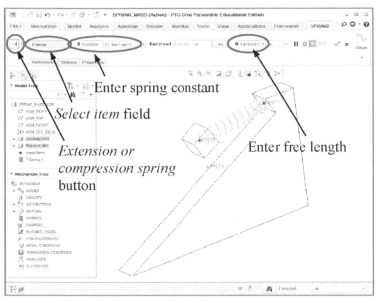

Figure 3-7 The *Spring Definition* Field

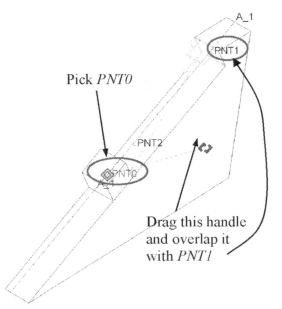

Figure 3-8 Defining the Spring

Figure 3-9 Adjust the Spring Diameter

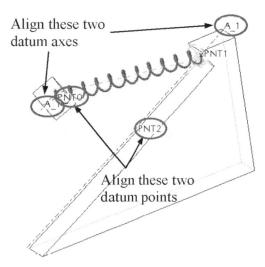

Figure 3-10 Align the Axes for Spring

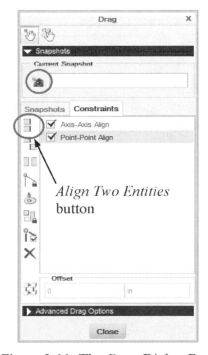

Figure 3-11 The *Drag* Dialog Box

Repeat the same process and pick two datum points, *PNT0* (block) and *PNT2* (ground), for aligning datum points. The block should be placed at its start position like that of Figure 3-4, which is 4 in. from the top of the slope; i.e., the spring is stretched 1 in. from its free length of 3 in.

Click the *Current Snapshots* button 📷 to create a snapshot *Snapshot1*, as shown in Figure 3-11.

Note that since the initial velocity is zero, we do not need to use the *Define Initial Conditions* button for any initial velocity. The motion model is now completely defined. We are ready to create and run a dynamic analysis.

Creating and Running a Dynamic Analysis

Click the *Mechanism Analysis* button 🗶 of the *Analysis* group to define an analysis. In the *Analysis Definition* dialog box (Figure 3-12) appearing, leave the default name, *AnalysisDefinition1*, and enter:

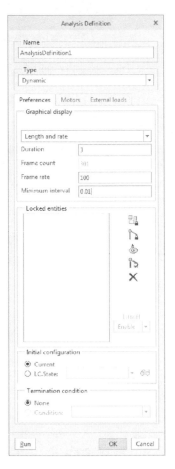

Type: *Dynamic*
Duration: *3*
Frame Rate: *100*
Minimum Interval: *0.01*
Initial Configuration: *Current*

Click *Run*. The progress of the analysis is shown at the bottom of the *Graphics* window, and the block starts sliding back and forth on the slope face. Click *OK* to save the analysis definition and close the dialog box.

Saving and Reviewing Results

Click the *Playback* button 🔄 of the *Analysis* group on top of the *Graphics* window (the same steps discussed in *Lesson 2*) to bring up the *Playbacks* dialog box and repeat the motion animation. On the *Playbacks* dialog box, click the *Save* button 💾 to save the results as a *.pbk* file.

Note that we want to create a measure to monitor the position of the block. We will choose the center point *PNT0* of the block and magnitude of the point position for the measure. To do so, click the *Measures* button ⬛ of the *Analysis* group on top of the *Graphics* window.

Figure 3-12 The *Analysis Definition* Dialog Box

In the *Measure Results* dialog box (Figure 3-13), click the *Create New Measure* button 🗋. The *Measure Definition* dialog box opens (Figure 3-14). Enter *Block_Position* for *Name*. Under *Type*, select *Position*. Select *PNT0* of the block for *Point or motion axis*. Leave *WCS* as the *Coordinate system* (default). Choose *Magnitude* for *Component* (default). Under *Evaluation Method*, leave *Each Time Step*. Click *OK* to accept the definition.

In the *Measure Results* dialog box (Figure 3-13) choose *AnalysisDefinition1* in the *Result Set* and click the *Graph* button 〰 on the top left corner to graph the measure.

The graph should be similar to that of Figure 3-15. Note that from the graph, the block will move along the slope face between 2 and 4 in. This is because the free length of the spring is 3 in. and we stretched the spring 1 in. to start the motion. Also, it takes about 0.75 seconds to complete a cycle (for example, the time interval between two consecutive peaks in Figure 3-15), which is large. This can be attributed to the fact that the spring is very soft (recall a small spring constant was defined). We will carry out hand calculations to verify these results. Before we do that, we will work on two more scenarios: with gravity and with the addition of an external force. Save your model before moving to the next scenario.

Figure 3-13 The *Measure Results* Dialog Box

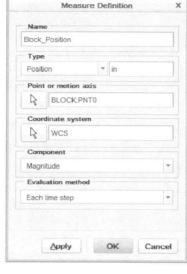

Figure 3-14 The *Measure Definition* Dialog Box

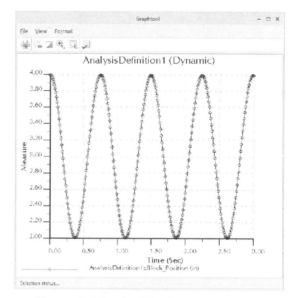

Figure 3-15 The *block* Position Graph

Scenario 2: With Gravity

We will add a gravitational acceleration and repeat the simulation. Do you expect to see any different results? Will the block move faster or slower due to gravity?

Click the *Gravity* button $\boxed{\downarrow g}$ of the *Properties and Conditions* group to bring up the *Gravity* dialog box (Figure 3-16, same as *Lesson 2*). Again, the default values of acceleration (386.088) and direction (0,–1,0) are what we need. No change is necessary. An arrow will appear in the *Graphics* window indicating the gravity.

Figure 3-16 The *Gravity* Dialog Box

Simply click *OK* in the *Gravity* dialog box. You will need to activate the gravity when you define a dynamic analysis.

Before running the analysis again, we will move the block back to its initial position, i.e., aligning *PNT0* and *PNT2*. This can be done by bringing back the snapshot *Snapshot1* saved earlier.

Click the *Drag Components* button 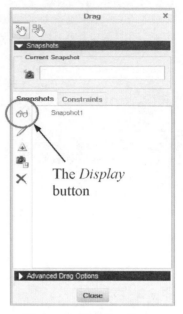 at the top of the *Graphics* window. In the *Drag* dialog box (Figure 3-17), the snapshot *Snapshot1* is listed. Choose the *Snapshot1* and click the *Display* button (the first on the left). The *block* should be placed back to the desired initial position. Close the *Drag* dialog box.

We will define a new dynamic analysis to include the gravity. Click the *Mechanism Analysis* button ⊠ to define a new analysis. In the *Analysis Definition* dialog box, use the default name *AnalysisDefinition2*; enter the same parameters as those of *Scenario 1*. Choose the *External Loads* tab, select *Enable Gravity*, and click *Run* (see Figure 3-18).

The block will start moving. It will move further down (see Figure 3-19) on the slope face due to gravity. Click *OK* to save the analysis definition. Note that you may graph the position of the block. The graph should be like that of Figure 3-20. In case you see something different, you may need to re-set the free length of the spring to 3 or set the initial configuration to *Snapshot1* using the *Drag Components* button.

Figure 3-17 The *Drag* Dialog Box

Note that from the graph, the *block* will move along the slope face between 4 and roughly 7.5 in. The vibration period is still about 0.75 seconds as it should be (why?). Save your model.

Scenario 3: With Gravity and External Force

In this scenario we will add an external force *p(t) = 10 cos 2t* at the center of the block in the downward direction along the slope. Note that the parameter in the cosine function of the external force must be converted from radian to degrees, as is assumed by *Mechanism*. Therefore, the force equation to enter becomes *p(t) = 10 cos (2t×180/π)*, i.e., *10 cos (114.59t)*.

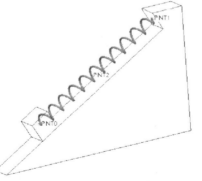

Figure 3-19 The *block* Motion of Scenario 2

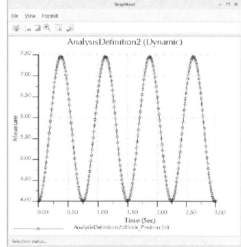

Figure 3-18 The *Analysis Definition* Dialog Box

Figure 3-20 The *block* Position Graph of Scenario 2

Note that *Mechanism* offers other types of functions for defining force and other motion entities, such as motor. More about functions can be found in Appendix D.

Click the *Force/Torque* button ⊢ of the *Insert* group; a new set of selections will appear at the top of the *Graphics* window for defining the force (see Figure 3-21). Choose *Translational motion* button (the first button from the left—should have been selected by default). Activate the *Select items* field by clicking it. Turn on the datum point display. Then, pick *PNT0* of the block from the *Graphics* window (or under the model tree). A force symbol will appear, pointing in the z-direction by default. Choose *User Defined* for *Function Type* (default is *Constant*) and click the *References* tab to define the force direction. Enter *–0.866*, *–0.5*, and *0* for *X*, *Y*, and *Z*, respectively (these values define a unit vector along the 30° slope); the force vector should point downward along the slope face in the *Graphics* window, as shown in Figure 3-21.

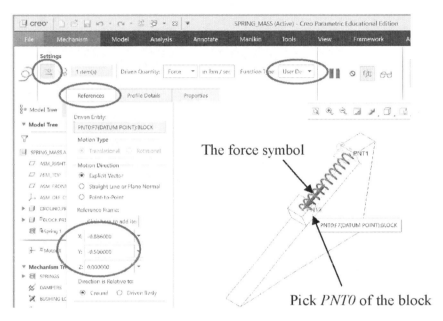

Figure 3-21 The *References* Tab for the Force or Torque Definition

Choose the *Profile Details* tab and click the *Add expression* button (second on the right; see Figure 3-22); a function *t* will appear in the left cell of the first row under the *Expression*. Enter *10*cos(114.59*t)* in the cell under *Expression*, and enter *0 < t < 6.28* in the cell under *Domain* for the lower and upper limits. As a result, the force will be applied for two complete cycles during the 6.28 second period. In this case, we will also specify the analysis duration as 6.28 (seconds). Click the graph button ⊠ (circled in Figure 3-22) to display the cosine function defined (see Figure 3-23). Click the *Properties* tab to enter *Force1* for name. Click the ✓ button on the right to accept the definition.

Create a new dynamic analysis *AnalysisDefinition3* by entering *6.28* for *Duration* and *100* for *Frame Rate*. Choose the *External Loads* tab and ensure *Force1* is listed under *Load* (Figure 3-24). If not, click the *Add new row* button (first of the three buttons to the right of the data cell circled in Figure 3-24) to add *Force1*. Make sure the *Enable Gravity* is selected and click *Run*.

The block will start moving. It will move further down along the slope due to both the external force and gravity.

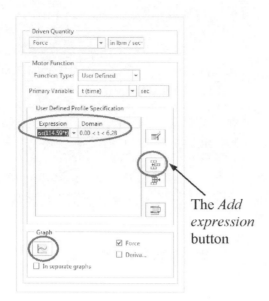

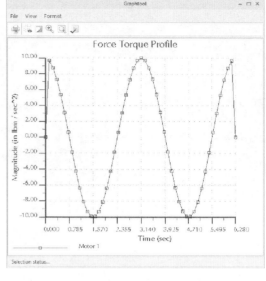

The *Add expression* button

Figure 3-22 The *Profile Details* Tab

Figure 3-23 The Force Function Graph

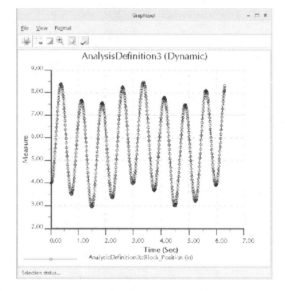

Figure 3-24 The *Analysis Definition* Dialog Box

Figure 3-25 The Block Position Graph of *Scenario 3*

Note that you may graph the position of the block. The graph should be like that of Figure 3-25. From the graph, the block travels along the slope for roughly 4.5 in., and the envelope of the motion amplitude varies. The result shows a different pattern compared to those of *Scenarios 1* and *2*. Does the result make sense to you? Save your model. We will verify the simulation results next.

3.4 Result Verifications

In this section, we will verify analysis results obtained from *Mechanism*.

There are two assumptions that we have to make in order to apply the particle dynamics theory to this spring-mass example:

(i)   The block is of a concentrated mass, and

(ii) No friction is present between the block and the slope face of the ground part.

We will start with the case of *Scenario 2* for equation of motion (i.e., with gravity), and then solve for both *Scenarios 1* and *2*. Note that *Scenario 1* is simply a special case of *Scenario 2* where gravity is turned off.

Equation of Motion: Scenarios 1 and 2

From the free-body diagram shown in Figure 3-26, applying Newton's Second Law and force equilibrium along the *x*-direction (i.e., the 30° slope), we have

$$\sum F_x = mg\sin\theta - k(x - U) = m\ddot{x} \qquad (3.1)$$

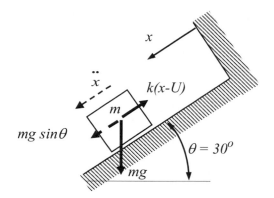

Therefore,

$$m\ddot{x} + k(x - U) = mg\sin\theta \qquad (3.2)$$

Figure 3-26 The Free-Body Diagram

where *m* is the mass of the block, *U* is the free length of the spring, *x* is the distance between the block and the top of the slope, measured from the top of the slope. The double dots on top of *x* represent the second derivative of *x* with respect to time. Rearrange Eq. 3.2; we have

$$m\ddot{x} + kx = mg\sin\theta + Uk \qquad (3.3)$$

where the terms on the right-hand side are constant.

This is a second-order ordinary differential equation. It is well known that the general solution of the differential equation is

$$x_g = A_1\cos\omega_n t + A_2\sin\omega_n t \qquad (3.4)$$

where $\omega_n = \sqrt{\dfrac{k}{m}}$, and A_1 and A_2 are constants to be determined with initial conditions. Note that the mass of the steel block is 0.2828 lb$_m$. This can be obtained from *Creo* by choosing the *Analysis* tab, then clicking the *Mass Properties* button. Therefore, $\omega_n = \sqrt{\dfrac{k}{m}} = \sqrt{\dfrac{20}{0.2828}} = 8.410$ rad/sec, and the natural frequency of the system is $f_n = \omega_n/2\pi = 1.338$ Hz. The period for a complete cycle $T = 1/f_n = 0.747$ seconds, which is close to that shown in all graphs (for example, see Figure 3-15).

The particular solution of Eq. 3.3 is

$$x_p = \frac{mg\sin\theta}{k} + U \qquad (3.5)$$

Therefore, the total solution is

$$x = x_g + x_p = A_1 \cos \omega_n t + A_2 \sin \omega_n t + \frac{mg \sin \theta}{k} + U \qquad (3.6)$$

The initial conditions for the spring-mass system are $x(0) = x_0 = 4$ in., and $\dot{x}(0) = 0$ in/sec. Plugging the initial conditions into Eq. 3.6 (you will have to take the derivative of Eq. 3.6 as well), we have

$$A_1 = x_0 - \left(\frac{mg \sin \theta}{k} + U \right), \text{ and}$$

$$A_2 = 0.$$

Hence, the overall solution is

$$x = \left(x_0 - \frac{mg \sin \theta}{k} - U \right) \cos \omega_n t + \frac{mg \sin \theta}{k} + U \qquad (3.7)$$

Note that Eq. 3.6 gives results for *Scenario 2*. For *Scenario 1*, simply set gravity g to 0; i.e.,

$$x = (x_0 - U) \cos \omega_n t + U \qquad (3.8)$$

Equations 3.7 and 3.8 can be implemented into *Microsoft Excel* shown in Figure 3-27.

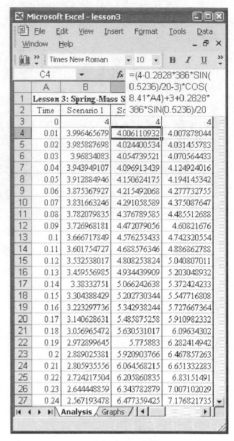

Figure 3-27 The *Excel* Spreadsheet

Note that Columns B and C in the spreadsheet show the results of Eqs. 3.8 and 3.7, respectively. Data in Column B are without gravity, i.e., Eq. 3.8 representing *Scenario 1*. Column C is for Eq. 3.7; i.e., *Scenario 2*.

Data in these two columns are graphed in Figures 3-28 and 3-29, respectively. Comparing Figures 3-28 and 3-29 with Figures 3-15 and 3-20, the results obtained from theory and *Mechanism* agree very well, which means the motion model has been properly defined, and *Mechanism* gives us good results.

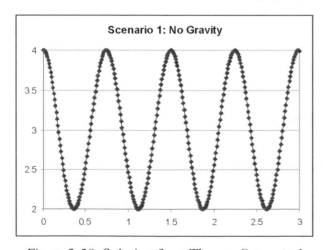

Figure 3-28 Solution from Theory: *Scenario 1*

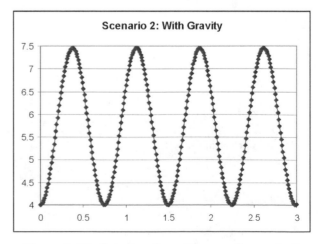

Figure 3-29 Solution from Theory: *Scenario 2*

Equation of Motion: Scenario 3

From the free-body diagram shown in Figure 3-26, for force equilibrium of *Scenario 3*, we must include the force $p = 10\ cos(2t)$ along the *X*-direction; i.e.,

$$m\ddot{x} + k(x - U) = mg\ sin\theta + f_0\ cos(\omega t) \tag{3.9}$$

where $f_0 = 10\ lb_m$ in/sec^2 and $\omega = 2$ rad/sec. Rearranging Eq. 3.9, we have

$$m\ddot{x} + kx = mg\ sin\theta + Uk + f_0\ cos(\omega t) \tag{3.10}$$

where the right-hand side consists of constant and time-dependent terms. For the constant term, the particular solution is identical to the previous case, i.e., Eq. 3.5. For the time-dependent term, i.e., $p = f_0\ cos(\omega t)$, the particular solution is

$$x_{p_2} = \frac{f_0}{k - \omega^2 m} cos\ \omega t \tag{3.11}$$

Therefore, the overall solution of Eq. 3.9 is

$$x = x_g + x_p + x_{p_2} = A_1\ cos\ \omega_n t + A_2\ sin\ \omega_n t + \frac{mg\ sin\theta}{k} + U + \frac{f_0}{k - \omega^2 m} cos\ \omega t \tag{3.12}$$

Plugging the initial conditions into the solution, we have

$$A_1 = x_0 - \left(\frac{mg\ sin\theta}{k} + U + \frac{f_0}{k - \omega^2 m}\right),\ \text{and}\ A_2 - 0.$$

Hence, the complete solution is

$$\begin{aligned}
x &= \left(x_0 - \frac{mg\ sin\theta}{k} - U - \frac{f_0}{k - \omega^2 m}\right) cos\ \omega_n t + \frac{mg\ sin\theta}{k} + U + \frac{f_0}{k - \omega^2 m} cos\ \omega t \\
&= \left[\left(x_0 - \frac{mg\ sin\theta}{k} - U\right) cos\ \omega_n t + \frac{mg\ sin\theta}{k} + U\right] + \frac{f_0}{k - \omega^2 m}\left(cos\ \omega t - cos\ \omega_n t\right)
\end{aligned} \tag{3.13}$$

Note that terms grouped in the first bracket of Eq. 3.13 are identical to those of Eq. 3.7, i.e., *Scenario 2*. The second term of Eq. 3.13 graphed in Figure 3-30 represents the contribution of the external force *p(t)* to the block motion. The graph shows that the amplitude of the block vibration should vary in time.

The overall solution of *Scenario 3*, i.e., Eq. 3.13, is a combination of graphs shown in Figures 3-29 and 3-30. In fact, Eq. 3.13 has been implemented in Column D of the spreadsheet. The data are graphed in Figure 3-31. As expected, the amplitude of the vibration is not constant. The vibration amplitude depends on time.

Comparing Figure 3-31 with Figure 3-25, the results obtained from theory and *Mechanism* are identical. Note that the spreadsheet shown in Figure 3-27 can be found at the publisher's website (filename: *lesson3.xls*).

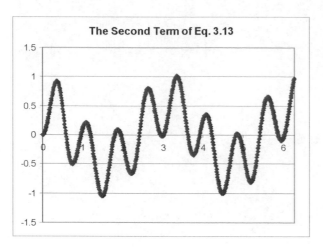

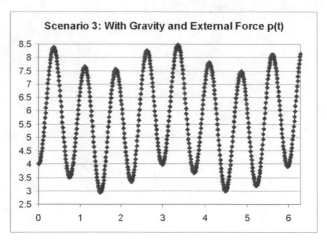

Figure 3-30 Graph of the Second Term of Eq. 3.13   Figure 3-31 Solution from Theory: *Scenario 3*

Exercises:

1.   Create and run a static analysis for Scenario 2. Where will the block be resting on the slope face due to gravity? Formulate a force equilibrium equation to solve for the resting position for the block. Is your result consistent with that of *Mechanism*?

2.   Show that Eq. 3.13 is the correct solution of *Scenario 3* governed by Eq. 3.9 by simply plugging Eq. 3.13 into Eq. 3.9.

3.   Repeat the *Scenario 3* of this lesson, except changing the external force to $p(t) = 10 \cos 8.41t$ lb_m in/sec^2. Will this external force change the vibration amplitude of the system? Can you simulate this resonance scenario in *Mechanism*?

4.   Add a damper with damping coefficient $C = 1$ lb_m/sec and repeat the *Scenario 1* simulation using *Mechanism*.

   (i)   Calculate the natural frequency of the system and compare your calculation with that of *Mechanism*.

   (ii) Derive and solve the equations that describe the position and velocity of the mass. Compare your solutions with those obtained from *Mechanism*.

Notes:

Lesson 4: A Simple Pendulum

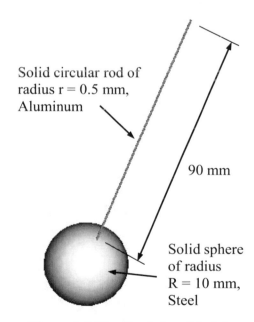

4.1 Overview of the Lesson

In this lesson, we will create a simple pendulum model using *Mechanism*. The pendulum will be released from a position slightly off the vertical line. The pendulum will then rotate freely due to gravity. In this lesson, we will learn how to create the pendulum motion model, run a dynamic analysis, and visualize the analysis results. The dynamic analysis results of the simple pendulum example can be verified using particle dynamics theory. Similar to *Lessons 2* and *3*, we will formulate the equation of motion; calculate the angular position, velocity, and acceleration of the pendulum; and compare our calculations with results obtained from *Mechanism*.

4.2 The Simple Pendulum Example

Physical Model

The physical model of the pendulum is composed of a sphere and a rod rigidly connected, as shown in Figure 4-1. The radius of the sphere is 10 mm. The length and radius of the thin rod are 90 mm and 0.5 mm, respectively. The top of the rod is connected to the wall with a pin joint. This pin joint allows the pendulum to rotate. The rod and sphere are made of aluminum and steel, respectively. Note that from the *Creo* material library the *AL2014* and *STEEL* material types have been selected for the rod and sphere, respectively. The *mmNs* system is selected for this example (millimeter for length, Newton for force, and second for time). In the *mmNs* unit system, the gravitational acceleration is 9,806 mm/sec^2.

Solid circular rod of radius r = 0.5 mm, Aluminum

90 mm

Solid sphere of radius R = 10 mm, Steel

The pendulum will be released from an angular position of 10 degrees from the vertical position along the pin joint. The rotation angle is intentionally kept small so that the particle dynamics theory can be applied to verify the simulation result.

Figure 4-1 The Pendulum Model

In this lesson, *Creo* parts of the pendulum example have been created for you. A partial assembly model with datum features required for this lesson is also provided. You can find the model files at the publisher's web site (www.sdcpublications.com). As mentioned in *Lesson 1*, datum features are extremely important in creating a successful motion model. In this lesson, we will use datum axis and datum points to define the pin joint. An additional datum axis will be used for defining the initial position of the pendulum. In this lesson, no solid part will be used for ground (unlike that of *Lesson 3*). The assembly datum features will be grouped for the ground body. Again, the datum coordinate system of the root assembly will be converted into the World Coordinate System (*WCS*) for the motion model.

Creo Parts and Assembly

The pendulum assembly consists of two parts, rod (*rod.prt*) and sphere (*sphere.prt*), as well as one assembly (*pendulum_partial.asm*) with datum features only. Note that you will need to open *pendulum_partial.asm* and assemble rod and sphere for a complete assembly. The exploded views shown in Figure 4-2 are from the complete assembly. They serve the purpose of illustration. In this lesson we will learn to create two new joints: the pin joint and the rigid joint. A pin joint allows rotation in one direction only. A rigid joint fixes all relative motion between parts. Obviously, a rigid joint will be defined between the rod and the sphere, and the pin joint will connect the top of the rod with the ground, allowing rotational motion along the *y*-direction of the assembly datum coordinate system *ASM_DEF_CSYS* (Figure 4-2b), which is *WCS*.

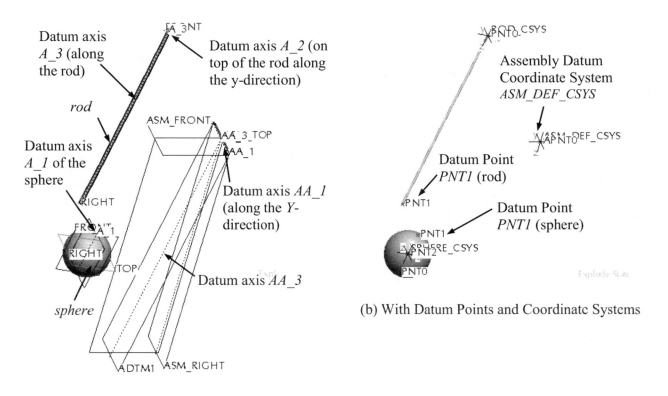

(a) With Datum Planes and Axes

(b) With Datum Points and Coordinate Systems

Figure 4-2 The Pendulum Assembly (Exploded View)

As shown in Figure 4-2a, the assembly *pendulum_partial.asm* consists of four datum planes. Note that the datum plane *ADTM1* was created by rotating *ASM_RIGHT* along the axis *AA_1* for a 10-degree angle, where the datum axis *AA_3* resides. The datum axis *AA_3* will be used to align with the datum axis *A_3* of the rod for defining the initial position of the pendulum. The datum axis *AA_1* aligns with the *y*-axis of the assembly datum coordinate system (*ASM_DEF_CSYS*). Note that the datum axis *AA_1* and *A_2* of the rod will be aligned to define the rotational axis of the pin joint. In addition, datum points *APNT0* and *PNT0* (rod) are used to restrain the translational movement of the rod as part of the pin joint. The rigid joint will be defined by aligning datum axes *A_1* of the sphere with *A_3* of the rod, and the datum points *PNT1* of the rod and *PNT1* of the sphere.

You may access the pendulum assembly by:

- Copying the *lesson4* folder that you downloaded from the publisher's web site to your computer;
- Starting *Creo* and changing the working directory to *lesson4* folder; and
- Opening the pendulum assembly by choosing, from the pull-down menu, *File > Open*.

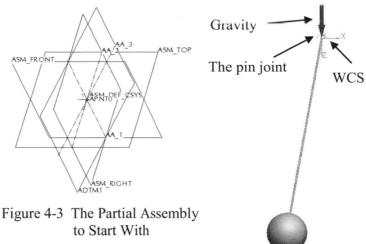

Figure 4-3 The Partial Assembly to Start With

Figure 4-4 The Motion Model

In the file open dialog box, choose *pendulum_partial.asm*. You should see, in the *Graphics* window, the partial pendulum assembly similar to the one shown in Figure 4-3.

Motion Model

As discussed, the rod and the sphere are connected by a rigid joint. Therefore, there is only one moveable body, which is free to rotate at the top end of the rod along the *y*-direction due to gravity. Note that the gravity is pointing in the positive *z*-direction. A complete view of the dynamic simulation model is given in Figure 4-4 shown in the *TOP* view (one of the saved views in the *Creo* model). Friction between the pendulum and the ground is assumed zero.

4.3 Using *Mechanism*

Creating a Complete Pendulum Assembly

Start *Creo*, set working directory, and open the assembly model: *pendulum_partial.asm* as discussed. Before you proceed further, you may want to make sure that the unit system is properly chosen for the parts and assembly. Check the unit system by choosing from the pull-down menu

File > Prepare > Model Properties.

In the *Model Properties* dialog box, *millimeter Newton Second (mmNs)* is listed, as shown in Figure 4-5. If this is not the case, click *change* for *Units* (Figure 4-5). The *Units Manager* dialog box will appear (Figure 4-6). In the *Units Manager* dialog box, choose *millimeter Newton second (mmNs)*, then click the *Set* button.

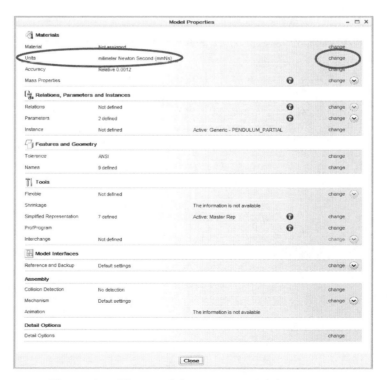

Figure 4-5 The *Model Properties* Dialog Box

In the *Changing Model Units* dialog box that appears (Figure 4-7), choose:

Interpret dimensions (for example 1" becomes 1mm), then click *OK*.

The red arrow should point to the unit system that you intend to use, as shown in Figure 4-6. Close the *Units Manager* dialog box, and then the *Model Properties* dialog box. You may want to open individual parts and check on their unit systems before you proceed.

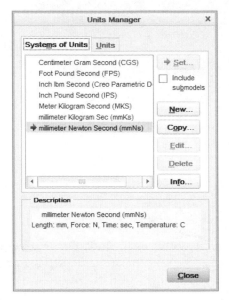

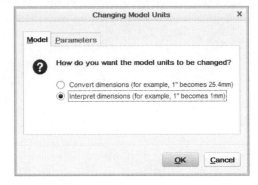

Figure 4-7 The *Changing Model Units* Dialog Box

Figure 4-6 The *Units Manager* Dialog Box

Now, we are ready to bring in the rod. Before we do that, keep in mind that we will define a pin joint by aligning *AA_1* (assembly) with *A_2* (rod) and *APNT0* (assembly) with *PNT0* (rod).

Click the *Assemble* button 🖼 and choose *rod.prt*. As soon as the rod appears, *Creo* assumes a configuration like the one shown in Figure 4-8a (you may see a different configuration), in which the datum features we are looking for are not quite visible. You may want to turn off the datum plane and the datum coordinate system display and move the rod away from the assembly datum feature, for example, to the location similar to the one shown in Figure 4-8b. You may turn off the *3D Dragger* (the red, blue, and green arrows and arcs) by clicking the *Show 3D Dragger* button ⊕ on top of the *Graphics* window. Zoom in the area as indicated in Figure 4-8b. Stay with the default view orientation for the time being.

In the *Component Placement* dashboard (upper left) choose the *Pin* joint (from *User Defined*), and pick *AA_1* (assembly) and *A_2* (rod) for defining the rotational axis of the pin joint, then pick *APNT0* (assembly) and *PNT0* (rod) to restrain the translational movement of the rod. You should see a pin joint symbol appears at the top of the rod like that of Figure 4-9.

Click the ✔ button to the right of the *Component Placement* dashboard to accept the definition.

Next, we will bring in the sphere and define a rigid joint to "glue" it to the end face of the rod. We will align axes *A_1* (rod) with *A_1* (sphere) and *PNT1* (rod) with *PNT1* (sphere).

Click the *Assemble* button and choose *sphere.prt*. In the *Component Placement* dashboard choose *Rigid* (from *User Defined*) and *Coincident* (to the right of the *Rigid* joint type), and pick *A_1* (rod) and *A_1* (sphere), then pick *PNT1* (rod) and *PNT1* (sphere). Note that no rigid joint symbol will be shown. The joint status (middle of the *Component Placement* dashboard) should be *Partially Constrained*. This is fine since the only free degree of freedom between the rod and the sphere is the rotation along the common axis (*A_1* axes in rod and sphere, respectively). Click the button at right of the *Component Placement* dashboard to accept the definition.

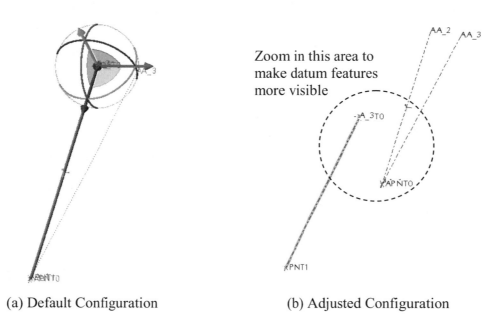

(a) Default Configuration   (b) Adjusted Configuration

Figure 4-8 Screen Captures of Rod Brought into Assembly

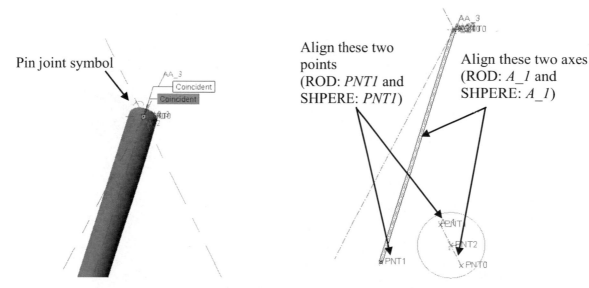

Figure 4-9 Pin Joint Defined for Rod   Figure 4-10 Defining a Rigid Joint

Creating a Dynamic Simulation Model

We now enter *Mechanism* by clicking the *Applications* tab on top of the *Graphics* window, and choosing *Mechanism* button .

We will define the gravitational acceleration and orient the rod to its initial configuration.

Click the *Gravity* button of the *Properties and Conditions* group at the top of the *Graphics* window. In the *Gravity* dialog box, the default value of acceleration (386.088) and default direction (0, 0, 1) appear. Keep the direction, change the *Magnitude* to 9806, and then click *OK*.

Click the *Drag Components* button , and the *Drag* dialog box appears (same as previous lessons). Click the *Constraints* tab and choose the *Align Two Entities* button (first on the left). Pick datum axes *A_1* of rod (or *A_1* of sphere) and *AA_3* (assembly). The pendulum should now align with the inclined datum axes *AA_3* that is 10 degree rotated from the vertical axes *AA_2* (or *z*-direction). Click the *Current Snapshots* button to create a snapshot (*Snapshot1*). You may want to choose the *TOP* view from the saved orientation list (click the *Saved Orientations* button of the *Graphics* toolbar at the top of the *Graphics* window) to see the pendulum model similar to that of Figure 4-11.

Creating and Running a Dynamic Analysis

Similar to previous lessons, we will create a dynamic simulation. Click the *Mechanism Analysis* button of the *Analysis* group. In the *Analysis Definition* dialog box, select *Dynamic* for *Type*. Leave the default name, *AnalysisDefinition1*. Enter:

Duration: *1.5*
Frame Rate: *100*
Minimum Interval: *0.01*

Click the *External Loads* tab, click *Enable Gravity*, and then click *Run*. You should see the pendulum start swinging back and forth along the pin joint in the *Graphics* window. Click *OK* to close the *Analysis Definition* dialog box.

Figure 4-11 The Pendulum Model

Saving and Reviewing Results

Click the *Playback* button of the *Analysis* group on top of the *Graphics* window (the same steps discussed in previous lessons) to bring up the *Playbacks* dialog box and repeat the motion animation. In the *Playbacks* dialog box, click the *Save* button to save the results as a *.pbk* file.

Note that we want to create a measure to monitor the position, velocity, and acceleration of the pendulum. We will choose the rotating axis of the pin joint for the measure. To do so, click the *Measures* button of the *Analysis* group on top of the *Graphics* window. In the *Measure Results* dialog box, click the *Create New Measure* button . The *Measure Definition* dialog box opens (Figure 4-12). Enter *angular_position* for *Name*. Under *Type*, select *Position*. Pick the pin joint symbol from the *Graphics* window for *Motion axis*. A double arrow symbol will appear (Figure 4-13), indicating that the rotational axis is selected. Note that the axis name listed under *Point or motion axis* may be different from what was listed in Figure 4-12. Leave *WCS* as the *Coordinate System* (default). Under *Evaluation Method*, leave *Each Time Step*. Click *OK* to accept the definition.

Figure 4-12 The *Measure Definition* Dialog Box

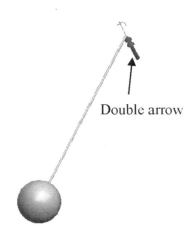

Figure 4-13 The Double Arrow Symbol

In the *Measure Results* dialog box choose *AnalysisDefinition1* in the *Result Set* and click the *Graph* button  at the top left corner to graph the measure. The graph should be similar to that of Figure 4-14, which is a sinusoidal function with amplitude between –10 and 10 degrees, as expected.

You may repeat the same steps to define angular velocity and angular acceleration measures for the pendulum (define the measures at the motion axis, i.e., the rotation axis of the pin joint). You should see graphs of the angular velocity and acceleration similar to those of Figures 4-15 and 4-16. Note that in this example the counterclockwise direction (as seen from the *FRONT* view) is positive as expected.

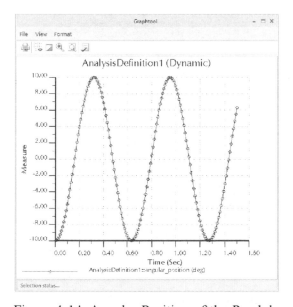

Figure 4-14 Angular Position of the Pendulum

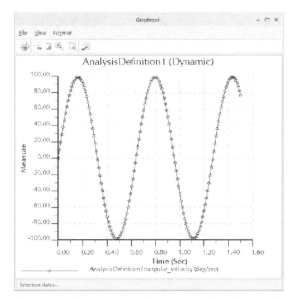

Figure 4-15 Angular Velocity of the Pendulum

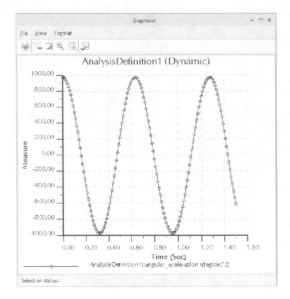

Figure 4-16 Angular Acceleration of the Pendulum

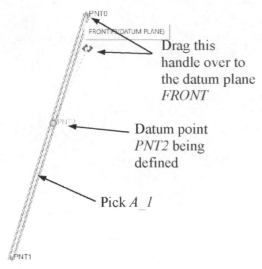

Figure 4-17 Creating New Datum Point

Create Additional Datum Point for Displaying Results

We want to also display the velocity magnitude at the rod mass center. Since there is no reference point (datum point) at the rod mass center, we will have to first create a datum point for the rod. The datum point will be created in the rod, along the axis A_1 45 mm from the top face of the rod (which is also the *FRONT* datum plane; see Figure 4-17).

To create a datum point, we will have to open the rod part model. The rod part model will be opened in a different window.

To create a datum point, you may click the *Point* button ⊞ from the *Datum* group on top of the *Graphics* window.

The *Datum Point* dialog box appears (see Figure 4-18) where the *References* field is active and waiting for you to make a selection. Pick datum axis A_1 in rod. A temporary datum point *PNT2* will appear in the rod with one handle (two small square brackets) aligning with the axis (see Figure 4-17), indicating that the datum point will stay on the axis.

In the *Datum Point* dialog box (Figure 4-18), the selected datum axis A_1 appears in the *References* field. Drag the dangling handle toward the datum plane on top, i.e., *FRONT*, until the datum plane label is highlighted, as shown in Figure 4-17. Then, release the mouse button. A dimension will appear in the rod indicating the distance of *PNT2* to the reference *FRONT*. Also in the *Datum Point* dialog box (Figure 4-18), an offset value appears with a number consistent to the dimension value in the *Graphics* window.

Figure 4-18 The *Datum Point* Dialog Box

Change the *Offset* value to 45 and click *OK*. The datum point *PNT2* is created. Be sure to turn on the datum point display (as well as datum point label) to see the datum point. Save the rod model and close the rod model window. Back to the pendulum assembly model, you should see that the newly created datum point *PNT2* appears in the rod.

Since there is no analysis result recorded for this newly created datum point, we will have to run the analysis again to generate analysis data for the datum point. Note that it is better to create datum points before the dynamic analysis to ensure you have complete results the first time.

Before you rerun the analysis, be sure to reset the assembly initial configuration using the snapshot *Snapshot1* saved earlier, where the axis *A_1* of the rod is aligned with *AA_3*. You may re-run the analysis by right-clicking the *AnalysisDefinition1 (DYNAMICS)* in the *Mechanism* model tree and choosing *Run*.

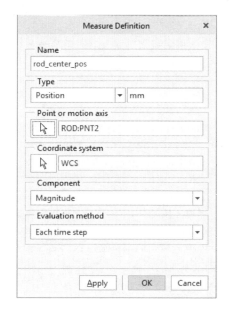

After rerunning the analysis, define a new measure at *PNT2*. In the *Measure Definition* dialog box (Figure 4-19), enter *rod_center_pos* for *Name*, pick *PNT2* from the *Graphics* window, and leave everything else as default. Click *OK*. Note that the *Component* field is left as *Magnitude*. If you graph this measure, what do you expect to see? (A straight horizontal line? Why?)

Choose the *rod_center_pos* measure from the *Measure Results* dialog box (Figure 4-20) and click the *Edit* button (right below the *New* on the left). In the *Measure Definition* dialog box, choose *X-component* in the *Component* field (that is, the horizontal direction), and click *OK*. Graph the measure. You should see a sinusoidal curve similar to that of Figure 4-21.

Figure 4-19 The *Measure Definition* Dialog Box

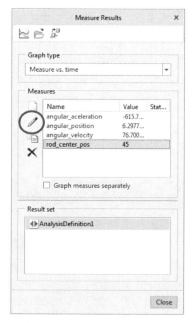

Figure 4-20 *Measure Results* Dialog Box

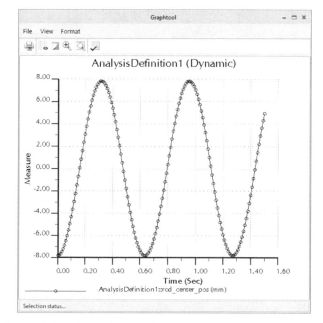

Figure 4-21 *X*-Position of Datum Point *PNT2* in Rod

Note that the pendulum is initially positioned to the left of the vertical axis. Therefore, the *X*-position of *PNT2* in the rod is on the negative side of the *X*-coordinate of the *WCS* (*ASM_DEF_CSYS*). The *X*-position of *PNT2* will vary between roughly –8 and 8 mm [or more precisely, ±7.814 mm; i.e., 45mm×cos(10°)], as shown in Figure 4-21.

4.4 Result Verifications

In this section, we will verify the analysis results obtained from *Mechanism* using particle dynamics theory.

There are four assumptions that we have to make in order to apply the particle dynamics theory to this simple pendulum problem:

(i)   Mass of the rod is negligible (this is why we chose *AL2014* for rod and *STEEL* for sphere),

(ii)   The sphere is of a concentrated mass,

(iii)   Rotation angle is small (remember the initial conditions we defined?), and

(iv)   No friction is present.

The pendulum model has been created to comply with these assumptions as much as possible. We expect that the particle dynamics theory will give us results close to those obtained through simulation. Two approaches will be presented to formulate the equations of motion for the pendulum: energy conservation and Newton's law.

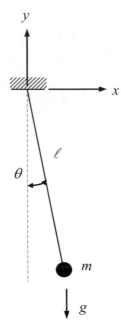

Figure 4-22 Particle Dynamics of Pendulum

Energy Conservation

Referring to Figure 4-22, the kinetic energy and potential energy of the pendulum can be written, respectively, as

$$T = \frac{1}{2} J \dot{\theta}^2 \qquad (4.1)$$

where *J* is the mass moment of inertia, i.e., $J = m\ell^2$;

and

$$U = mg\ell (1 - \cos \theta) \qquad (4.2)$$

According to the energy conservation theory, the total mechanical energy, which is the sum of the kinetic energy and potential energy, is a constant with respect to time; i.e.,

$$\frac{d}{dt}(T + U) = 0 \qquad (4.3)$$

where *t* represents time. Hence

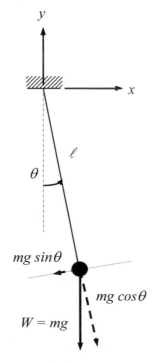

Figure 4-23 Free Body Diagram

$$\frac{d}{dt}\left(\frac{1}{2}m\ell^2\,\dot{\theta}^2 + mg\ell(1-\cos\theta)\right) = m\ell^2\,\ddot{\theta} + mg\ell\sin\theta = 0 \tag{4.4}$$

Therefore,

$$\ddot{\theta} + \frac{g}{\ell}\sin\theta = 0,\ \text{and}$$

$$\ddot{\theta} + \frac{g}{\ell}\theta = 0 \tag{4.5}$$

when $\theta \approx 0$.

Newton's Law

From the free-body diagram shown in Figure 4-23, the equilibrium equation of moment at the origin along the z-direction (perpendicular to the paper) can be written as:

$$\sum M = -mg\ell\sin\theta = J\ddot{\theta} = m\ell^2\,\ddot{\theta} \tag{4.6}$$

Hence

$$\ddot{\theta} + \frac{g}{\ell}\sin\theta = 0,\ \text{and}$$

$$\ddot{\theta} + \frac{g}{\ell}\theta = 0 \tag{4.7}$$

when $\theta \approx 0$.

Note that the same equation of motion has been derived from two different approaches. The linear ordinary second-order differential equation can be solved analytically.

Solving the Differential Equation

It is well known that the solution of the differential equation is

$$\theta = A_1\cos\omega_n t + A_2\sin\omega_n t \tag{4.8}$$

where $\omega_n = \sqrt{\dfrac{g}{\ell}}$, and A_1 and A_2 are constants to be determined by initial conditions. Note that

$\omega_n = \sqrt{\dfrac{g}{\ell}} = \sqrt{\dfrac{9806}{100}} = 9.903$ rad/sec, and the natural frequency of the system is $f_n = \omega_n/2\pi = 1.576$ Hz. The time period for a complete cycle is $T = 1/f_n = 0.634$ seconds, which is very close to that shown in all graphs (for example, Figure 4-14).

The initial conditions for the pendulum are $\theta(0) = \theta_0 = -10$ degrees, and $\dot{\theta}(0) = 0$ degree/sec. Plugging the initial conditions into the solution, we have

$A_1 = \theta_0 = -10$ degrees, and $A_2 = 0$.

Hence, the solutions are

$$\theta = \theta_0 \cos \omega_n t \qquad\qquad\qquad\qquad\qquad\qquad\qquad\qquad\qquad\qquad (4.9a)$$

$$\dot{\theta} = -\theta_0 \omega_n \sin \omega_n t \qquad\qquad\qquad\qquad\qquad\qquad\qquad\qquad\qquad (4.9b)$$

$$\ddot{\theta} = -\theta_0 \omega_n^2 \cos \omega_n t \qquad\qquad\qquad\qquad\qquad\qquad\qquad\qquad\qquad (4.9c)$$

The above equations represent angular position, velocity, and acceleration of the pin joint. These equations can be implemented into, for example, *Microsoft Excel* shown in Figure 4-24, for numerical solutions. Columns B, C, and D in the spreadsheet show the results of Eqs. 4.9a, b, and c, respectively, between 0 and 1.5 seconds with an increment of 0.005 seconds. Data in these three columns are graphed in Figures 4-25, 26, and 27, respectively. Comparing Figures 4-25 to 27 with Figures 4-14 to 16, the results obtained from theory and simulation are very close, which means the motion model has been properly defined, and *Mechanism* gives us good results.

However, these results are not completely identical. This is because the *Mechanism* model is not really a simple pendulum since mass of the rod is non-zero. If you reduce the diameter of the rod, the *Mechanism* results should approach those obtained through theory calculations.

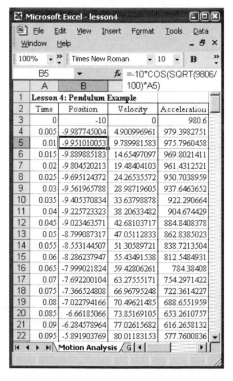

Figure 4-24 The *Excel* Spreadsheet

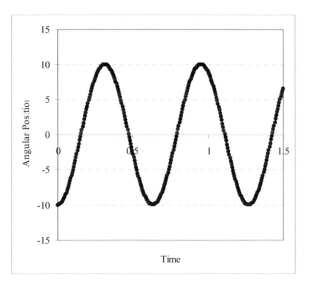

Figure 4-25 Angular Position from Theory

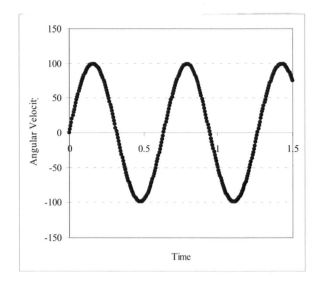

Figure 4-26 Angular Velocity from Theory

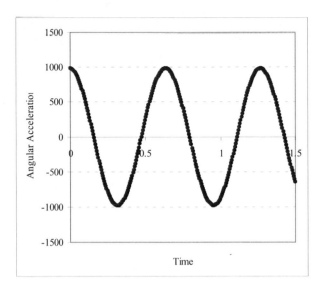

Figure 4-27 Angular Acceleration from Theory

Exercises:

1.   Define a measure and graph the maximum angular velocity of the pendulum. Also, calculate the maximum angular velocity of the pendulum from theory and compare your results with those obtained from *Mechanism*.

2.   Create a spring-damper-mass system, as shown in Figure E4-1, using *Mechanism*. Note that the unstretched spring length is 3 in. The radius of the ball is 0.5 in. and the material is steel (mass density: $0.2637 \ lb_m/in^3$).

      (i)   Find the spring length in the equilibrium condition using *Mechanism*. Hint: define a *Static* analysis to simulate the equilibrium position of the mass.

      (ii)   Solve the same problem using Newton's laws. Compare your results with those obtained from *Mechanism*.

3.   If a force $p = 2 \ lb_f$ is applied to the ball as shown in Figure E4-1, repeat both (i) and (ii) of Problem 2.

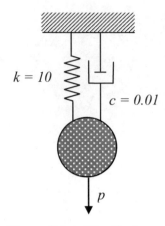

Figure E4-1 The Spring-
Mass-Damper System

Lesson 5: A Slider-Crank Mechanism

5.1 Overview of the Lesson

In this lesson you will learn how to create simulation models for a slider-crank mechanism and carry out three analyses. First, we will conduct a position analysis and turn on interference checking to see if parts collide. It is important to make sure no interference exists between parts while the mechanism is in motion. We will also learn to create a trace curve to trace the trajectory of a selected point in the mechanism in a position analysis. The second analysis is kinematic. A kinematic analysis will calculate velocity and acceleration data instead of just positional data obtained from position analysis. The final analysis will be dynamic, where we will add a firing force to the piston for a dynamic analysis. This lesson will start with a brief overview about the slider-crank assembly created in *Creo*. More joint types will be introduced. You will learn how to select proper joint types together with placement constraints to connect parts. At the end, we will verify the kinematic analysis results using theory and computational methods commonly found in *Mechanism Design* textbooks.

5.2 The Slider-Crank Example

Physical Model

The slider-crank mechanism shown in Figure 5-1 is essentially a four-bar linkage. They are commonly found in mechanical systems, e.g., internal combustion engines and oil-well drilling equipment. For the internal combustion engine, the mechanism is driven by a firing load that pushes the piston (or slider), converting the reciprocal motion into rotational motion at the crank.

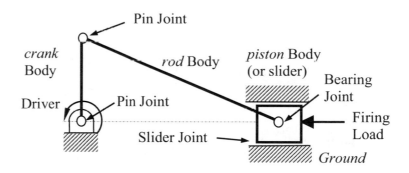

Figure 5-1 Schematic View of the Slider-Crank Mechanism

In the oil-well drilling equipment, a torque is applied at the crank. The rotational motion is converted to a reciprocal motion at the slider (or piston) that digs into the ground. Note that in any case the length of the crank must be smaller than that of the rod in order to allow the mechanism to operate (Grashof's law).

Note that the unit system chosen for this example is *IPS* (in-lb$_f$-sec). No friction is assumed between any pair of bodies.

Creo Parts and Assembly

The slider-crank system consists of four parts, crank (*crank.prt*), rod (*rod.prt*), pin (*pin.prt*), and piston (*piston.prt*), as shown in the exploded view in Figure 5-2. Figure 5-2 also shows datum points in both parts and assembly. Datum points (and datum axes) created in parts (and subassemblies) will be used to define connections (joints) between bodies as well as serve as the application points of forces.

In this lesson, we will start with a partial assembly, *slider_crank_partial.asm*. The partial assembly consists of datum planes, datum axes, and datum points, as shown in Figure 5-3. Again, these assembly datum features will be converted into ground body. Note that the datum axes *AA_1* and *AA_2* and datum point *APNT0* will be used for creating joints—specifically, the pin joint between the ground and the crank, and the slider joint between the ground and the piston. Note that datum axis *AA_1* is offset 0.4 in. along the negative z-direction of the assembly coordinate system *ASM_DEF_CSYS*. This is necessary for creating the slider joint between the piston and the ground since we would like to have the translational direction of the slider joint line up with the x-axis.

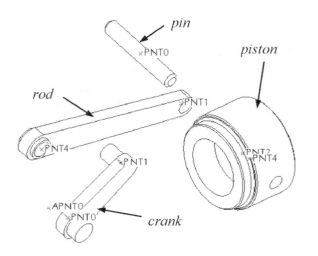

Figure 5-2 Slider-Crank Assembly (Exploded View)

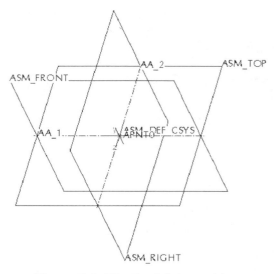

Figure 5-3 The Partial Assembly

Motion Model

In this example, we will define two pin joints, one bearing joint, and one slider joint. The first pin joint (*Pin1*) allows rotational motion between the crank and the ground body, as shown in Figure 5-4. The second pin joint (*Pin2*) will be created to allow rotation motion between the rod and the crank. After assembling the crank and the rod, the system should have two degrees of freedom, allowing the crank and rod to rotate at their respective pin joints independently.

Next, the pin part will be assembled to the rod rigidly using placement constraints, still maintaining two dof's. Then, the piston will be assembled to the pin by defining a bearing joint (aligning axis *A_1* in pin part with datum point *PNT2* in piston). The piston is free to translate along the axis *A_1* (pin) and rotate in all three directions. The total dof's will now increase to six.

Finally, the *piston* will be assembled to the ground body by defining a slider joint. The slider joint will be created by aligning two parallel axes (*A_1* in piston and *AA_1* in the assembly) and two datum planes (*DTM3* in piston and *ASM_TOP* in the assembly). The slider joint will allow only one translational movement between the piston and the assembly (ground), i.e., along the common axes without rotation, therefore forcing the bearing joint between the piston and pin part to behave more like a pin joint, allowing only rotation along the common axes. The slider-crank mechanism is now restricted to planar motion, with three rotations (*Pin1*, *Pin2*, and *Bearing1*), and one translation (*Slider1*) motion. However, all three rotations and the translation motion are related to form a closed loop mechanism, leaving only one free dof, which can be any one of the rotations or the translation of the piston.

The total number of degrees of freedom of the slider-crank mechanism can also be calculated as follows:

3 (bodies) × 6 (dof's/body) – 2 (pins) × 5 (dof's/pin) – 1 (slider joint) × 5 (dof's/slider) – 1 (bearing joint) × 2 (dof's/bearing) = 18 – 17 = 1

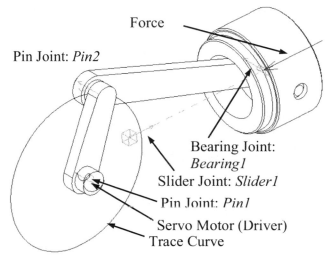

Although this calculation gives consistent results as discussed earlier for the single-piston engine shown in Figure 1-12 of *Lesson 1*, the way the joints are chosen for a motion model is not unique. One would probably create three pin joints (replacing the bearing joint with a pin joint between the piston and the pin) and one slider joint, which still leads to a physically meaningful motion model. However, the total dof's will become –2. This is because there are three redundant dof's created in the system. This is fine since *Mechanism* filters out redundant dof's and excludes them in analysis. You may want to check the redundancy following steps shown in Appendix A.

Figure 5-4 The Motion Model

For this slider-crank mechanism there is only one dof remaining, which means one single driver or force will move the mechanism and uniquely determine the configuration of the mechanism in time domain. Joints defined in this simulation model are summarized in Table 5-1. The pairs of datum points and datum axes created in the parts and assembly for defining these four joints can be seen in the top and front views of the mechanism, as shown in Figure 5-5.

To go through this lesson, you need to download the *lesson5* folder from the publisher's website to your hard drive. Before going forward, please spend a few minutes to review the parts and assembly, especially paying close attention to the datum axes and datum points.

Table 5-1 Joints Defined in the Simulation Model

	Ground Body	crank	rod/pin	piston
crank	*Pin1* A_1 (crank)/AA_2 and *PNT0/APNT0*		*Pin2* A_2 (crank)/A_1 (rod) and flat faces	
rod/pin		*Pin2* A_2 (crank)/A_1 (rod) and flat faces		*Bearing1* PNT2 (piston)/A_1 (pin)
piston	*Slider1* A_1 (piston)/AA_1 and *DTM3* (piston)/ *ASM_TOP*		*Bearing1* PNT2 (piston)/A_1 (pin)	

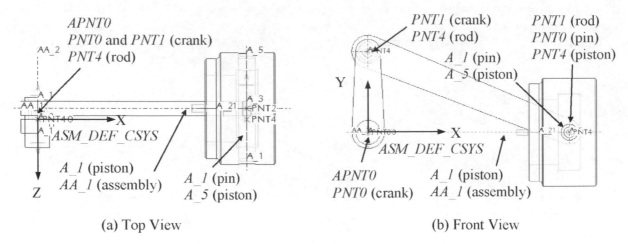

(a) Top View   (b) Front View

Figure 5-5 Locations of Datum Points and Datum Axes

We need to also define a servo motor (driver) that will rotate the crank through the rotation axis of the joint *Pin1* (between the crank and the ground body). As a result, the motor will drive the slider-crank mechanism. The driver will rotate the crank at a constant angular velocity of 360 degrees/sec. We will first conduct a position analysis and check interferences. Then, we will carry out a kinematic analysis for the slider-crank mechanism and later add a force for a dynamic analysis.

5.3 Using *Mechanism*

Creating an Assembly

In the *lesson5* folder, you should see the following files:

lesson5\slider_crank_partial.asm
lesson5\crank.prt
lesson5\rod.prt
lesson5\pin.prt
lesson5\piston.prt

Start *Creo*, select working directory, and open the assembly model: *slider_crank_partial.asm*. You should see an assembly with numerous datum features, as shown in Figure 5-2. Before we proceed, you may want to make sure that the unit system is properly chosen for the part. The unit system for all four parts and the assembly should be *IPS* (in-lb$_f$-sec).

Now, we are ready to bring in the crank. Before we do that, please note that we will define a pin joint by aligning axis *A_1* (crank) with *AA_2* (assembly) and *PNT0* (crank) with *APNT0* (assembly). The pin joint will allow one rotation dof along the z-direction of the assembly coordinate system.

Click the *Assemble* button [icon] and choose *crank.prt*. As soon as the crank appears, *Creo* assumes a configuration like the one shown in Figure 5-6a (only show datum axes and datum points). You may want to turn off the datum plane and datum coordinate system display to show only datum axes and datum points.

In the *Component Placement* dashboard, choose *Pin* joint from the *User Defined* list, and pick *A_1* (crank) and *AA_2* (assembly). The crank will be repositioned to a configuration where *A_1* aligns with *AA_2*. Next, pick *PNT0* (crank) and *APNT0* (assembly). A pin joint symbol appears (Figure 5-6b). You may want to use the *No Hidden* option to unshade the crank in order to see the pin joint symbol. Even though you could drag/rotate the crank, it is recommended that you stay with the current configuration and not drag the component. Click the ✅ button to accept the definition.

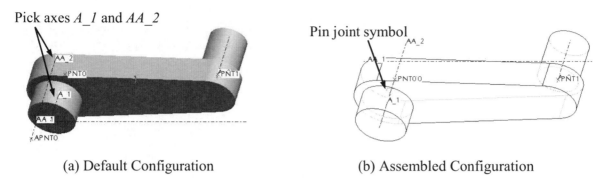

(a) Default Configuration   (b) Assembled Configuration

Figure 5-6 Assembling Crank to the Ground Body

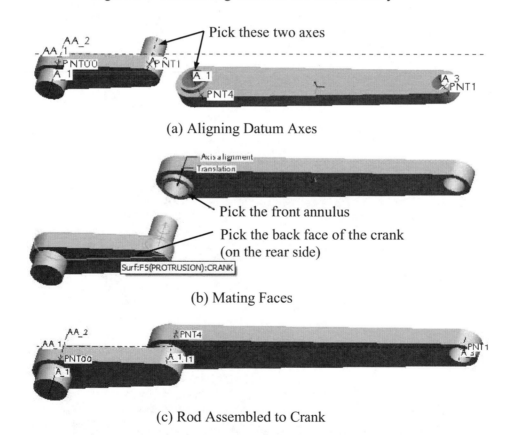

(a) Aligning Datum Axes

(b) Mating Faces

(c) Rod Assembled to Crank

Figure 5-7 Assembling Rod to Crank

Next, we will bring in the rod and define a pin joint to connect it to the small end of the crank. We will align axes *A_1* (rod) with *A_2* (crank) and coincide the front annulus of the rod with the back face of the crank. The rod will be shown in a default configuration, like that of Figure 5-7a when it is brought into the assembly. Use the *3D Dragger* to move and orient the rod to a configuration similar to that of Figure 5-7b. Make sure *PNT4* (rod) is on the front side of the rod. In the *Component Placement* dashboard from the *Joint Type* list choose the *Pin* joint, and pick *A_1* (rod) and *A_2* (crank). Use the *Move* option to translate the rod away from the crank like the one shown in Figure 5-7b if necessary. Pick the front annulus of the rod and the back face of the crank (see Figure 5-7b), and you should see that the rod is properly assembled to the crank (Figure 5-7c). Click the ✔ button to accept the definition.

The next part we bring in is the pin (*pin.prt*). Since the pin is connected to the rod rigidly, we will assemble the pin using standard placement constraints. We will align axes *A_1* (pin) with *A_3* (the small end of the rod) and *DTM2* (pin) with *DTM2* (rod), as shown in Figures 5-8a and 5-8b. The rod will be brought in with a default configuration like that of Figure 5-8a. In the *Component Placement* dashboard click *Placement* tab then choose *Coincident* from the *Constraint Type* (see Figure 5-9). Pick *A_1* (pin) and *A_3* (rod). Click *New Constraint* and choose *Coincident*. Pick *DTM2* (pin) and *DTM2* (rod) shown in Figure 5-8b, and you should see that the pin is properly assembled to the crank (Figure 5-8c). The status at the top of the *Graphics* window (see Figure 5-9) indicates that the part is fully constrained. In fact, the pin is allowed to rotate along its axis *A_1*. The status of *fully constrained* is due to the fact that the *Allow Assumptions* is checked (see Figure 5-9). Click the ✔ button to accept the definition.

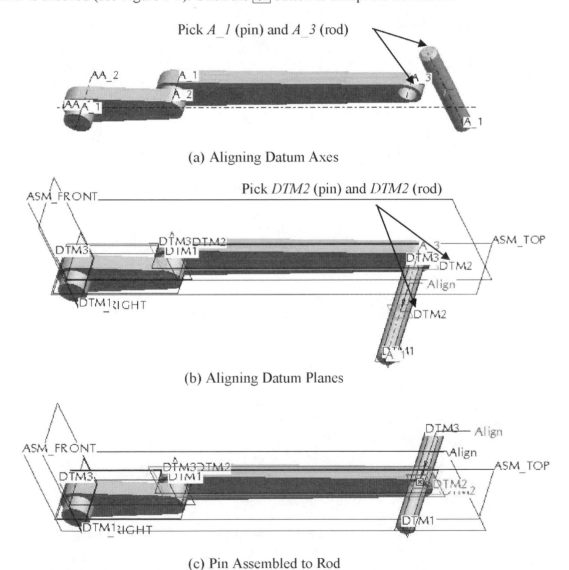

(a) Aligning Datum Axes

(b) Aligning Datum Planes

(c) Pin Assembled to Rod

Figure 5-8 Assembling Pin to Rod

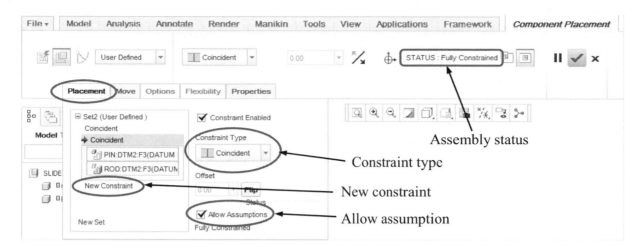

Figure 5-9 Placement Constraint Dashboard

The final part we are bringing in is the piston. The piston will be assembled to the pin part and the ground body using a bearing joint and a slider joint, respectively. The piston will be shown in a default configuration when it is brought into the assembly, similar to that of Figure 5-10a. In the *Component Placement* dashboard, click *Bearing* from the *User Defined* list. Pick *A_1* (pin) and *PNT2* (piston). *PNT2* is now allowed to move along axis *A_1* and rotate in all three directions, therefore a bearing joint. Next, we will define a slider joint.

Click the *Placement* tab in the *Component Placement* dashboard; a bearing joint is now listed, as shown in Figure 5-10b. Click the *New Set* button (lower left corner) to create a new joint. Click the new joint (for example, *Connecting_4* in Figure 5-10c) appearing in the *Component Placement* dashboard, and choose *Slider* for *Set Type*. Pick *A_1* (piston) and *AA_1* (assembly), and *DTM3* (piston) and *ASM_TOP* (assembly), as shown in Figure 5-10d. Note that some datum planes were hidden in Figure 5-10d for better illustration. Note that it may be easier to pick these datum entities from the model tree. You should see that the piston is properly assembled and a slider joint appears. Click the ✔ button to accept the definition.

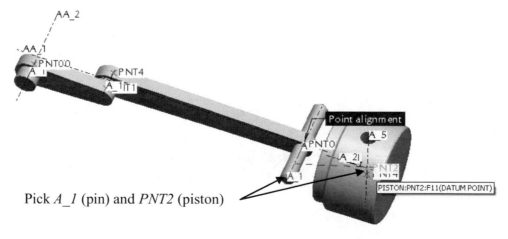

(a) Aligning Datum Point to Axis for Bearing Joint

Figure 5-10 Assembling Piston to Ground Body and Pin

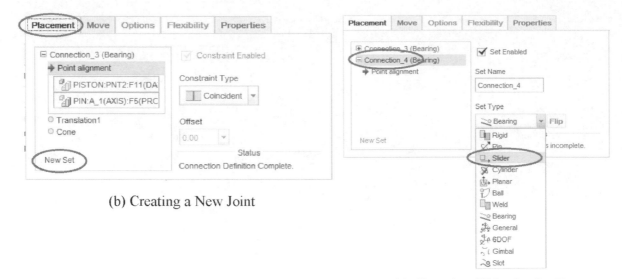

(b) Creating a New Joint

(c) Choosing Slider for Set Type

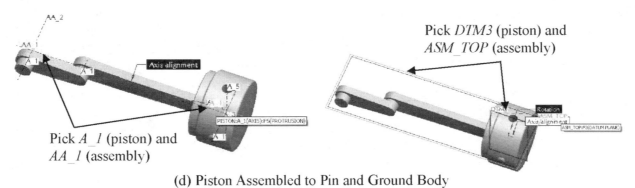

Pick *DTM3* (piston) and
ASM_TOP (assembly)

Pick *A_1* (piston) and
AA_1 (assembly)

(d) Piston Assembled to Pin and Ground Body

Figure 5-10 Assembling Piston to Ground Body and Pin (cont'd)

Now, we have completely assembled the mechanism. You may want to use the *Drag* option to move the mechanism. Click the *Drag Components* button ⬚ of the *Component* group at the top of the *Graphics* window. Then, click the crank and move the mouse. You should see the slider-crank mechanism moves as expected (Figure 5-11).

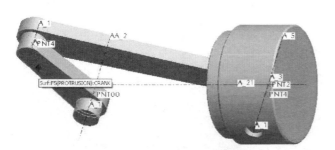

Figure 5-11 Move the Mechanism by
Dragging the Crank

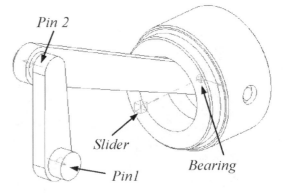

Pin 2

Slider

Bearing

Pin1

Figure 5-12 Joints Defined for the Slider-
Crank Mechanism

Creating Simulation Models

We are now ready to enter *Mechanism*. To enter *Mechanism*, simply click the *Applications* tab on top of the *Graphics* window, and choose *Mechanism* button 🔧. When you enter *Mechanism*, please review joint symbols to make sure you have two pin joints (*Pin1* between the crack and the assembly, *Pin2* between the rod and the crank), one bearing joint (between the piston and the pin), and one slide joint, as shown in Figure 5-13.

We will first define an initial condition for the simulation model. The initial condition will be defined similar to the sketch shown in Figure 5-1, where the crank is pointing vertically upward. Then, we will create a driver at joint *Pin1* to conduct both position and kinematic analyses. Two datum planes will be aligned for the initial configuration, *DTM1* (crank) and *ASM_TOP* (assembly), as shown in Figure 5-13a. We will use the *Drag* option to align the planes and create a snapshot of the configuration for future use.

Click the *Drag Components* button 👆 at the top of the *Graphics* window, and the *Drag* dialog box appears. Click the *Constraints* tab and choose the *Align Two Entities* button (first on the left). Choose two datum planes, *DTM1* (crank) and *ASM_TOP* (assembly), as shown in Figure 5-13a. The crank should now sit upright as that of Figure 5-13b. Click the *Current Snapshot* button 📷 on top, and a default name *Snapshot1* will appear. Click *Close*. The snapshot *Snapshot1* has been saved for future use. Before starting analysis, make sure you bring up this snapshot as the initial configuration for analysis.

Pick *DTM1* (crank) and *ASM_TOP* (assembly)

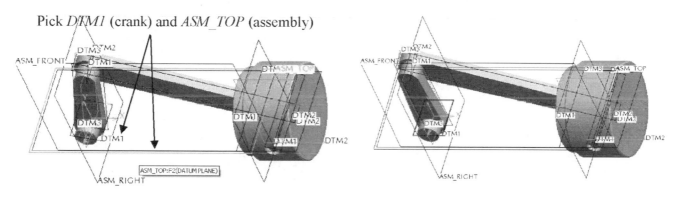

(a) Selecting Two Datum Planes for Alignment   (b) Initial Configuration

Figure 5-13 Defining Initial Configuration

Now we create a driver at the rotational axis of the pin joint between the crank and the ground body (*Pin1*). The driver rotates the crank at a constant angular velocity of 360 degrees/sec.

Click *Servo Motors* 🔘 of the *Insert* group on top of the *Graphics* window; a new set of selections will appear (see Figure 5-14). Pick *Pin1* from the *Graphics* window between the crank and the ground (see Figure 5-12 to locate *Pin1*). After picking the pin joint, larger arrows appear to confirm your selection. Note that you may want to turn off all datum features in order to see the pin joint. You may need to click the *Flip* button to make sure you pick the correct direction of the rotation axis. The axis must point in a direction like that of Figure 5-14. Also note the changes in the selections above the *Graphics* window; for example, the rotational motion button 🔄 is now selected, and *Driven Quantity* changed to *Angular Position*. Select *Angular Velocity* for *Driven Quantity*. Click the *References* tab to make sure that *Connection_1.first_rot_axis* has been selected for *Driven Entity*, as circled in Figure 5-14.

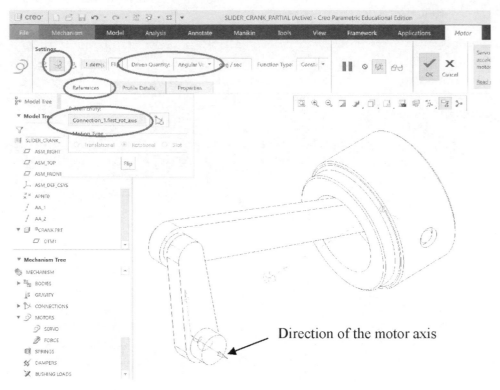

Figure 5-14 Defining a Servo Motor

The next step is to specify the profile of the motor. Click the *Profile Details* tab (next to the *References* tab shown in Figure 5-14), choose *Angular Velocity* for *Driven Quantity* and *Constant* (default) for *Function Type* (see Figure 5-15). Enter *360* for the constant *A* under *Coefficients*.

Click the *Properties* tab to enter *Motor1* for name. Click the *OK* ✔ button on the right to accept the definition. A motor symbol appears in the *Graphics* window overlapping with *Pin1*, as circled in Figure 5-16.

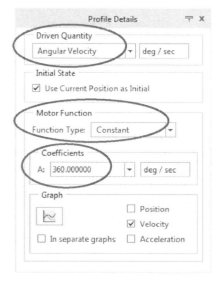

Figure 5-15 The *Profile Details* Tab

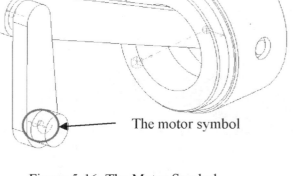

Figure 5-16 The Motor Symbol

Creating and Running a Position Analysis

Position analysis is also called *Repeated Assembly* analysis in *Mechanism*. It is a series of assembly analyses driven by servo motors. A position analysis simulates the mechanism's motion, satisfying the requirements of the servo motors profiles and all joints (and cam-follower, slot-follower, or gear-pair connections), and records position data for the mechanism's various components. It does not take force and mass into account in performing the analysis. Therefore, you do not have to specify mass properties for your mechanism. Dynamic entities in the model, such as springs, dampers, gravity, forces/torques, and force motors, do not affect a position analysis. A position analysis gives you positions of components over time, trace curves of the mechanism's motion, and interference between components.

Click the *Mechanism Analysis* button 📐 to define an analysis. In the *Analysis Definition* dialog box that appears, enter:

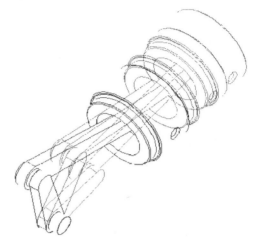

Name: *Position_Analysis*
Type: *Position*
Start Time: *0*
End Time: *1*
Frame Rate: *100*
Minimum Interval: *0.01*
Initial Configuration: *Current*

Click *Run* button. In the *Graphics* window, the mechanism starts moving. The crank rotates 360 degrees as expected (Figure 5-17). Click *OK* to save the analysis definition.

Figure 5-17 Animation

Saving and Reviewing Results

Click the *Playback* button 🔁 to bring up the *Playbacks* dialog box and repeat the motion animation. On the *Playbacks* dialog box, click the *Save* button 💾 to save the results as a *.pbk* file.

Note that we want to create a measure to monitor the position of the piston. We will choose the center point *PNT2* of the piston and magnitude of the point position for the measure. To do so, click the *Measures* button 📈. In the *Measure Results* dialog box appearing, click the *Create New Measure* button 🗋. The *Measure Definition* dialog box opens (Figure 5-18). Enter *Piston_Position* for *Name*. Under *Type*, select *Position*. Pick *PNT2* in the *piston* part (you may need to zoom into the center of the piston and right click the overlapping datum entity labels a few times until *PNT2* appears). Leave *WCS* as the *Coordinate System* (default). Choose *Magnitude* as the *Component* (default). Under *Evaluation Method*, leave *Each Time Step*. Click *OK* to accept the definition.

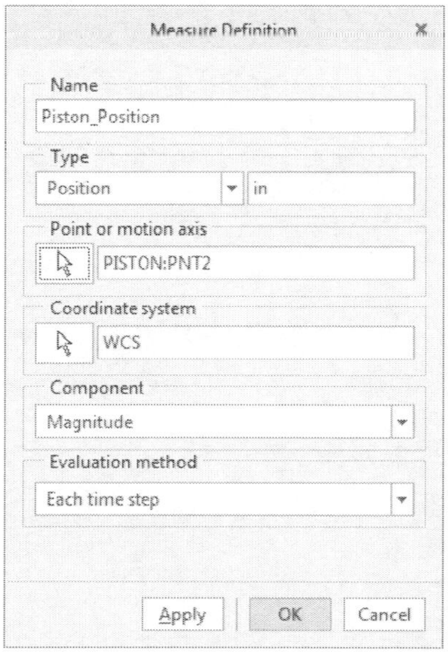

Figure 5-18 The *Measure Definition* Dialog Box

In the *Measure Results* dialog box, choose *Position_Analysis* in the *Result Set* and click the *Graph* button ⊵ on the top left corner to graph the measure.

The graph should be similar to that of Figure 5-19. Note that from the graph, the piston moves between about 7.5 and 11 in. horizontally, in reference to the *WCS*. At the starting point, the crank is in the vertical position, and the piston is in fact at 7.42 in. (that is, $\sqrt{8^2 - 3^2}$) to the right of *WCS*. Note that the lengths of the crank and rod are 3 and 8 in., respectively. When the crank rotates to 90 degrees counterclockwise, the position becomes 5 (which is 5 = 8–3). When the crank rotates 270 degrees, the piston position is 11 (which is 11 = 8+3).

Repeat the same steps to define a measure for the angular position of the pin joint *Pin1*. The graph of the angular position of *Pin1* should be similar to that of Figure 5-20. The graph shows a straight line between 90 and 450 degrees within the one-second analysis period. Certainly, this is due to the fact that a driver of a constant angular velocity is employed to drive the mechanism.

Now we will create a trace curve to trace the location of pin joint *Pin2* (between crank and rod). A trace curve graphically represents the motion of a point or vertex relative to a part in the mechanism. This is especially useful for designing a cam profile. What will the trace curve look like for this case?

Click *Analysis* above the *Graphics* window (see Figure 5-21), and choose *Trace Curve*; the *Trace Curve* dialog box appears (Figure 5-22). Pick *SLIDER_CRANK_PARTIAL.ASM* from the model tree for *Paper Part* (that is, the reference), and *PNT1* (crank) for the trace point (under *Point, vertex, or curve endpoint* in the *Trace Curve* dialog box). Leave *2D* for *Curve Type*, choose *Position_Analysis*, and click *Preview* (or *OK*). A circle centered at *APNT0* and passing through *PNT1* appears (Figure 5-23). This circle specifies the trace of point *PNT1* with respect to the ground body (*slider_crank_partial.asm*).

Interference Checking

Next we learn how to perform interference checking. In *Mechanism*, interference occurs when two parts collide, indicating that one part is interfering with another part's ability to move. You should check for interference if you are modeling the detailed geometry of the mechanism, like this slider-crank example. The interference checking is straightforward in *Mechanism*. The capability is built in playbacks.

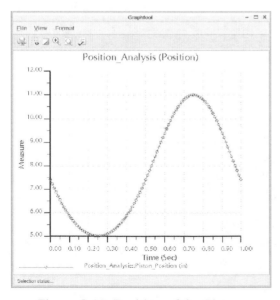

Figure 5-19 Position of the Piston

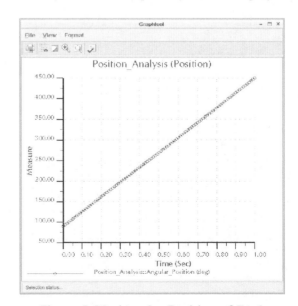

Figure 5-20 Angular Position of *Pin1*

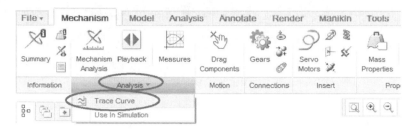

Figure 5-21 Selecting *Trace Curve* Under *Analysis* Group

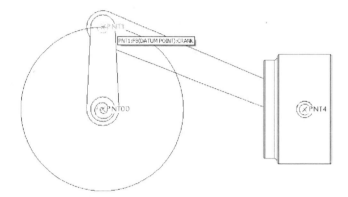

Figure 5-22 The *Trace Curve*
Dialog Box

Figure 5-23 The Trace Curve of Datum Point *PNT1* (crank)

Click the *Playback* button ◀▶ of the *Analysis* group on top of the *Graphics* window to bring up the *Playbacks* dialog box (Figure 5-24). Click the *Collision Detection Settings* button. In the *Collision Detection Settings* dialog box (Figure 5-25), click *Global Collision Detection*, and then click *OK*. The *Global* option allows *Mechanism* to check for interference between all parts in the mechanism. You may choose to sound a warning or stop the animation when interference is detected. The *No Collision Detection* option (default) will turn off interference checking.

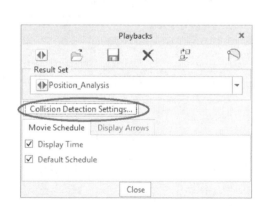

Figure 5-24 The *Playbacks* Dialog Box

Figure 5-25 The *Collision Detection
Settings* Dialog Box

From the *Playbacks* dialog box, click the *Play Current Result Set* button ◀▶ (top left) to bring up the *Animate* dialog box. Note that it may take a few seconds for *Mechanism* to calculate interference at each time frame. Click the *Play* button from the *Animate* dialog box to see the animation again. In the

Graphics window, you should see that a small portion of the inner surface of the piston and the cylindrical surface at the right end of the rod is highlighted in red (Figure 5-26), meaning interference is detected between these two parts. You may choose *FRONT* view from the *Saved Orientations* button [icon] of the *Graphics* toolbox to see the model similar to that of Figure 5-26b.

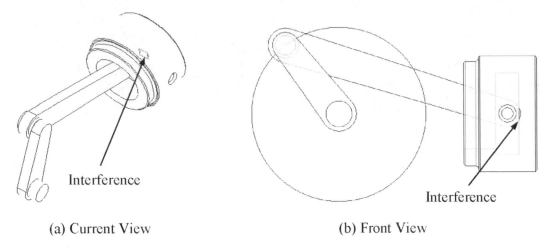

(a) Current View   (b) Front View

Figure 5-26 Interference Checking

Creating and Running a Kinematic Analysis

We will define a kinematic analysis for the same slider-crank simulation model. The definition includes:

Name: *Kinematic_Analysis*
Type: *Kinematic*
Start Time: *0*
End: *1*
Frame Rate: *100*
Minimum Interval: *0.01*
Initial Configuration: *Current*

The mechanism should be in its initial configuration; i.e., crank is in the upright position. If not, click the *Drag Components* button [icon] at the top of the *Graphics* window, and the *Drag* dialog box appears (Figure 5-27).

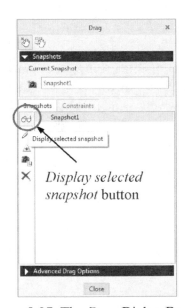

Figure 5-27 The *Drag* Dialog Box

Choose *Snapshot1* and click the *Display selected snapshot* button (first on the left). The mechanism will be restored to its initial configuration.

Run the kinematic analysis. In the *Graphics* window, the mechanism starts moving. The crank rotates 360 degrees just like the position analysis. Moreover, velocity and acceleration data are calculated. Define additional measures, for example, angular velocity of *Pin2*, and piston velocity in the *X*-direction (at *PNT2* of piston). Graphs of the measures should be like those of Figures 5-28 and 5-29, respectively.

Accuracy of the result graphs obtained from *Mechanism* will be verified later in this lesson.

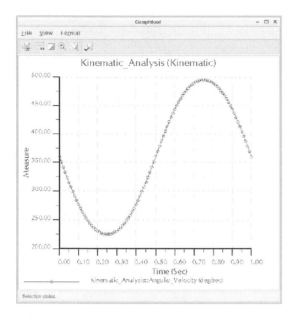

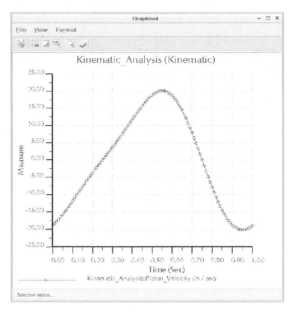

Figure 5-28 Angular Velocity of Pin Joint: *Pin2*

Figure 5-29 Piston Velocity in *X*-Direction (*PNT2*)

Creating and Running a Dynamic Analysis

A force mimicking the engine firing load (acting along the negative *X*-direction of *ASM_DEF_CSYS*) will be added to the piston for a dynamic analysis. It will be more realistic if the force can be applied when the piston starts moving left and the force can only be applied for a short period of time. In order to do so, we will have to define measures that monitor the position of the piston for the firing load to be activated. Unfortunately, such a capability is not yet available in *Mechanism*. Therefore, the force is simplified as an impulse force with a triangular shape that has a peak of 1,000 lb$_f$ at 0.5 seconds, as shown in Figure 5-30. The force will be applied over a one-second period. This force will be defined as a point force at datum point *PNT2* of the piston (Figure 5-31).

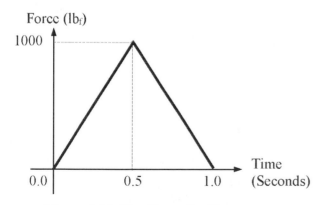

Figure 5-30 The Force Profile

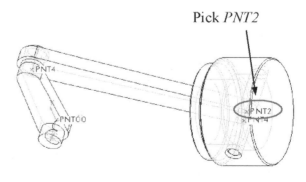

Figure 5-31 Pick Point for Force Application

Click the *Force/Torque* button [⊩] of the *Insert* group; a new set of selections will appear at the top of the *Graphics* window for defining the force (see Figure 5-32). Choose *Translational motion* button [⟳] (the first button from the left, which should have been selected by default). Activate the *Select items* field by clicking it. Turn on the datum point display. Then, pick *PNT2* of the piston from the *Graphics* window (or pick from the model tree). Note that you may want to turn off all datum features displayed except for datum points (see Figure 5-31). After picking the point, a larger arrow appears pointing along a default direction (Z-direction: 0, 0, 1). We change it to (−1, 0, 0) by clicking the *References* tab, and entering (−1, 0, 0) for (X, Y, Z), as shown in Figure 5-32.

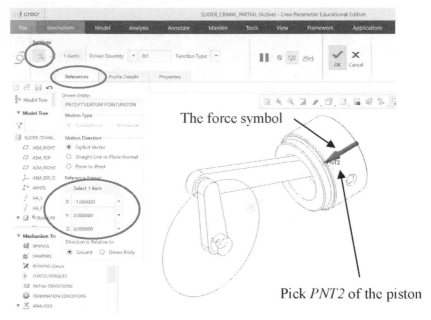

The force symbol

Pick *PNT2* of the piston

Figure 5-32 The *References* Tab for the Force or Torque Definition

Next, we enter the load magnitude as a function as described in Figure 5-30. Click the *Profile Details* tab and choose *Table* for *Function Type* (see Figure 5-33). Click the *Add rows to table* button (first of the two buttons on the right) three times to create three rows. Enter three pairs of data (0, 0), (0.5, 1,000), and (1, 0), and click the *Graph* button ⬚ to show the graph of the force profile (Figure 5-34). Note that you must make sure that the unit system is in-lb$_f$-sec. If you are still using the default unit system, in-lb$_m$-sec, you will have to enter a peak force 386,000 lb$_m$ in/sec^2 at 0.5 second (instead of 1,000 lb$_f$).

Click the *Properties* tab and enter *Firing_load* for *Name*. Click the ☑ button on the right to accept the definition. A force symbol should appear at *PNT2* in the piston in the *Graphics* window (Figure 5-35).

Now we are ready for defining a dynamic analysis. Before we proceed, we will reset the initial position. Click the *Drag Components* button 🖑. In the *Drag* dialog box, click *Snapshot1*, and click the *Display selected snapshot* button (first on the left). The mechanism will be reset to the configuration where the crank is in the upright position.

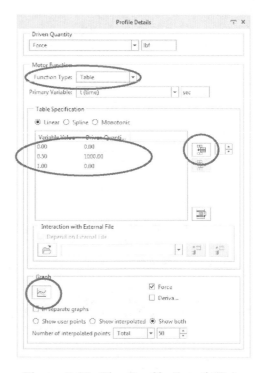

Figure 5-33 The *Profile Details* Tab

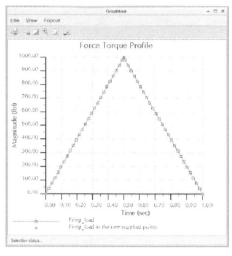

Figure 5-34 Graph of the force
profile

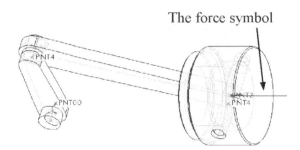

Figure 5-35 Force Symbol at Piston

Click the *Mechanism Analysis* button ⊠ to define an analysis. In the *Analysis Definition* dialog box, enter:

Name: *Dynamic_Analysis*
Type: *Dynamic*
Duration: *3*
Frame Rate: *100*
Minimum Interval: *0.01*
Initial Configuration: *Current*

We will have to delete the motor to comply with our scenario. To do so, click the *Motors* tab, select *MOTOR1*, then click the *Delete highlighted row(s)* button (2nd on the right). Click the *External Loads* tab just to make sure the *Firing_load* is included. If not, click the *Add rows to table* button (first of the two buttons on the right) to include the firing load. Click *Run*. In the *Graphics* window, the mechanism starts moving. The crank rotates about one and a half cycles. Click *OK* to save the analysis definition.

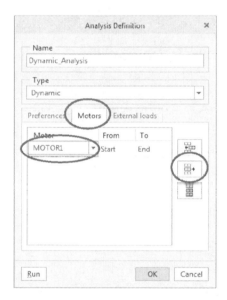

Figure 5-36 The *Motors* Tab of
the *Analysis Definition* Dialog Box

You may also see graphs for some of the measures, e.g., piston position as shown in Figure 5-37. Do not forget to save your model.

5.4 Result Verifications

In this section, we will verify the kinematic analysis results using computational method found in *Mechanism Design* textbooks. Recall that in kinematic analysis, position, velocity, and acceleration of given points or axes in the mechanism are analyzed.

In kinematic analysis, forces and torques are not involved. All bodies (or links) are assumed massless. Hence, mass properties defined for bodies are not influencing the analysis results.

The slider-crank mechanism is a planar kinematic analysis problem. A vector plot that represents the positions of joints of the planar mechanism is shown in Figure 5-38. The vector plot serves as the first step in computing position, velocity, and accelerations of the mechanism.

The position equations of the system can be described by the following vector summation,

$$Z_1 + Z_2 = Z_3 \qquad (5.1)$$

where

$$Z_1 = Z_1 \cos \theta_A + i Z_1 \sin \theta_A = Z_1 e^{i\theta_A}$$
$$Z_2 = Z_2 \cos \theta_B + i Z_2 \sin \theta_B = Z_2 e^{i\theta_B}$$
$$Z_3 = Z_3, \text{ since } \theta_C \text{ is always } 0.$$

The real and imaginary parts of Eq. 5.1, corresponding to the X and Y components of the vectors, can be written as

$$Z_1 \cos \theta_A + Z_2 \cos \theta_B = Z_3 \qquad (5.2a)$$
$$Z_1 \sin \theta_A + Z_2 \sin \theta_B = 0 \qquad (5.2b)$$

In Eqs. 5.2a and 5.2b, Z_1, Z_2, and θ_A are given. We are solving for Z_3 and θ_B. Equations 5.2a and 5.2b are non-linear functions of Z_3 and θ_B. Solving them directly for Z_3 and θ_B is not straightforward. Instead, we will calculate Z_3 first, using trigonometric relations; i.e.,

$$Z_2^2 = Z_1^2 + Z_3^2 - 2Z_1 Z_3 \cos \theta_A$$

Hence,

$$Z_3^2 - 2Z_1 \cos \theta_A Z_3 + Z_1^2 - Z_2^2 = 0$$

Solving Z_3 from the above quadratic equation, we have

$$Z_3 = \frac{2Z_1 \cos \theta_A \pm \sqrt{(2Z_1 \cos \theta_A)^2 - 4(Z_1^2 - Z_2^2)}}{2} \qquad (5.3)$$

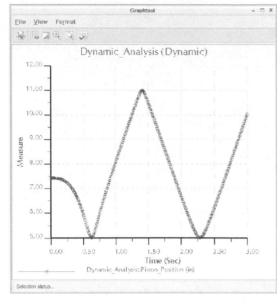

Figure 5-37 Piston Position of Dynamic Simulation

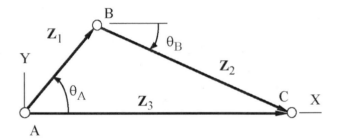

Figure 5-38 Vector Plot of the *slider-crank* Mechanism

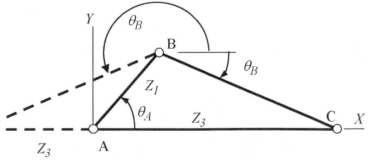

Figure 5-39 Two Possible Configurations

where two solutions of Z_3 represent the two possible configurations of the mechanism shown in Figure 5-39. Note that point C can be either at C or C' for given Z_1 and θ_A.

From Eq. 5.2b, θ_B can be solved by

$$\theta_B = sin^{-1}\left(\frac{-Z_1 \sin\theta_A}{Z_2}\right) \tag{5.4}$$

Similarly, θ_B has two possible solutions, corresponding to vector $\mathbf{Z}_3$.

Taking derivatives of Eqs. 5.2a and 5.2b with respect to time, we have

$$-Z_1 \sin\theta_A\,\dot\theta_A - Z_2 \sin\theta_B\,\dot\theta_B = \dot Z_3 \tag{5.5a}$$

$$Z_1 \cos\theta_A\,\dot\theta_A + Z_2 \cos\theta_B\,\dot\theta_B = 0 \tag{5.5b}$$

where $\dot\theta_A = \dfrac{d\theta_A}{dt} = \omega_A$ is the angular velocity of the driver, which is a constant. Note that Eqs. 5.5a and 5.5b are linear functions of $\dot Z_3$ and $\dot\theta_B$. Rewrite the equations in a matrix form; i.e.,

$$\begin{bmatrix} Z_2 \sin\theta_B & 1 \\ Z_2 \cos\theta_B & 0 \end{bmatrix}\begin{bmatrix} \dot\theta_B \\ \dot Z_3 \end{bmatrix} = \begin{bmatrix} -Z_1 \sin\theta_A\,\dot\theta_A \\ -Z_1 \cos\theta_A\,\dot\theta_A \end{bmatrix} \tag{5.6}$$

Equation 5.6 can be solved by

$$\begin{bmatrix} \dot\theta_B \\ \dot Z_3 \end{bmatrix} = \begin{bmatrix} Z_2 \sin\theta_B & 1 \\ Z_2 \cos\theta_B & 0 \end{bmatrix}^{-1}\begin{bmatrix} -Z_1 \sin\theta_A\,\dot\theta_A \\ -Z_1 \cos\theta_A\,\dot\theta_A \end{bmatrix} = \frac{1}{-Z_2 \cos\theta_B}\begin{bmatrix} 0 & -1 \\ -Z_2 \cos\theta_B & Z_2 \sin\theta_B \end{bmatrix}\begin{bmatrix} -Z_1 \sin\theta_A\,\dot\theta_A \\ -Z_1 \cos\theta_A\,\dot\theta_A \end{bmatrix}$$

$$= \frac{1}{-Z_2 \cos\theta_B}\begin{bmatrix} Z_1 \cos\theta_A\,\dot\theta_A \\ Z_1 Z_2 \cos\theta_B \sin\theta_A\,\dot\theta_A - Z_1 Z_2 \sin\theta_B \cos\theta_A\,\dot\theta_A \end{bmatrix}$$

$$= \begin{bmatrix} -\dfrac{Z_1 \cos\theta_A\,\dot\theta_A}{Z_2 \cos\theta_B} \\[4mm] -\dfrac{Z_1\left(\cos\theta_B \sin\theta_A\,\dot\theta_A - \sin\theta_B \cos\theta_A\,\dot\theta_A\right)}{\cos\theta_B} \end{bmatrix} \tag{5.7}$$

Hence

$$\dot{\theta}_B = -\frac{Z_1 \cos\theta_A \dot{\theta}_A}{Z_2 \cos\theta_B} \tag{5.8}$$

and

$$\dot{Z}_3 = Z_1\left(\tan\theta_B \cos\theta_A \dot{\theta}_A - \sin\theta_A \dot{\theta}_A\right) \tag{5.9}$$

In this example, $Z_1 = 3$, $Z_2 = 8$, and the initial conditions are $\theta_A(0) = \pi/2$ and $\theta_B(0) = \sin^{-1}(3/8)$.

The solutions can be implemented using a spreadsheet. The *Excel* file, *lesson5.xls*, can be found at the publisher's website. As shown in Figure 5-40, Columns A to I represent time, Z_1, Z_2, $\dot{\theta}_A$, θ_A, Z_3, θ_B, $\dot{Z}_3$, and $\dot{\theta}_B$, respectively. Note that in this calculation, $Z_3(0)>0$ is assumed, hence $\theta_B(0)<0$, as illustrated in Figure 5-39. This is consistent with the initial conditions we defined for the motion model.

Figures 5-41 to 5-43 show the graphs of data in Columns F, G, and H. Comparing Figures 5-41 and 5-42 with Figures 5-19 and 5-29, the motion analysis results agree with those of the spreadsheet calculations.

Note that in Figures 5-28 and 5-43 the graphs are different in sign. This is because the direction of the pin joint, *Pin2*, is pointing in the negative Z-direction in *Mechanism*, which is different from what we assumed in our calculations. Also, the angle units are different.

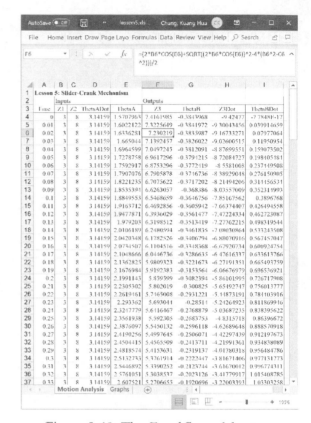

Figure 5-40 The *Excel* Spreadsheet

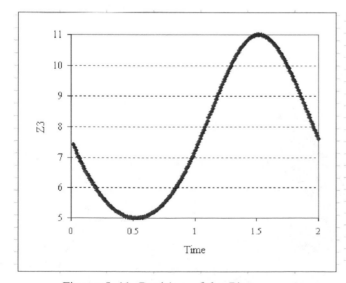

Figure 5-41 Position of the Piston

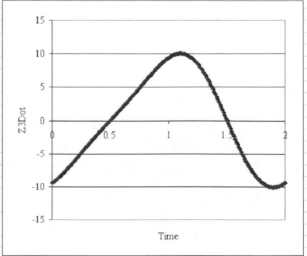

Figure 5-42 Velocity of the Piston

Note that the accelerations of a given joint in the mechanism can be formulated by taking one more time derivative of Eqs. 5.5a and 5.5b. The resulting two linear equations can be solved using *Excel*. This is left as an exercise.

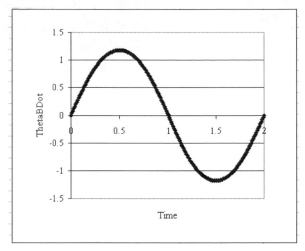

Figure 5-43 Angular Velocity of the
Pin Joint: *Pin2*

Exercises:

1. Derive the acceleration equations for the slider-crank mechanism by taking derivatives of Eqs. 5.5a and 5.5b with respect to time. Solve these equations for the linear acceleration of the piston and the angular acceleration of the pin joint *Pin2* using a spreadsheet. Compare your solutions with those obtained from *Mechanism*.

2. Use the same slider-crank model to conduct a static analysis using *Mechanism*. The static analysis in *Mechanism* should give you equilibrium configuration(s) of the mechanism due to gravity. Show the equilibrium configuration(s) of the mechanism and use the energy method you learned from *Statics* to verify the equilibrium configuration(s).

3. Change the length of the crank from 3 to 5 in. in *Creo*. Repeat the kinematic analysis discussed in this lesson. In addition, change the crank length in the spreadsheet (*Microsoft Excel* file, *lesson5.xls*). Generate position and velocity graphs from both *Mechanism* and the spreadsheet. Do they agree with each other? Is the maximum slider velocity increase due to a longer crank? Is there interference occurring in the mechanism?

4. Download four *Creo* parts from the publisher's website to your computer (folder name: *Exercise 5-4*).

    (i)   Use these four parts, i.e., crankshaft, connecting rod, piston pin, and piston (see Figure E5-1), to create an assembly like the one shown in Figure E5-2. Note that the crankshaft must be 45° CCW.

    (ii) Create a motion model for kinematic analysis. Conduct motion analysis by defining a driver that drives the crankshaft at a constant angular speed of 1,000 rpm.

    (iii) Use the spreadsheet *lesson5.xls* to calculate the piston velocity. Compare your calculations with those obtained from *Mechanism*.

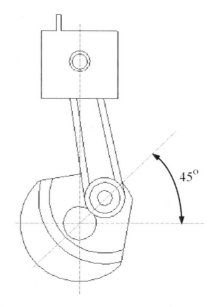

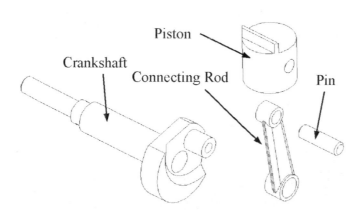

Figure E5-1 Four *Creo* Parts   Figure E5-2 Assembled Configuration

Lesson 6: A Compound Spur Gear Train

6.1 Overview of the Lesson

In this lesson we will discuss how to simulate motion of a spur gear train. A gear train is a set or system of gears arranged to transfer torque and power from one part of a mechanical system to another. A gear train consists of a driving gear that is attached to the input shaft, a driven gear attached to the output shaft, and idler gears that interpose between the driving and driven gears in order to maintain a desired direction of the output shaft or to increase the distance between the driving and driven gears. There are different kinds of gear trains, such as simple gear train, compound gear train, epicylic gear train, etc., depending on their functionality and how the gears are arranged. The gear train we are simulating in this lesson is a compound spur gear train, in which two or more gear pairs are used to transmit torque and power. All gears included in this lesson are spur gears; therefore, the shafts that these gears are mounted on are parallel to the longitudinal direction of the teeth on the spur gears.

In *Mechanism*, each gear in a gear train comprises one body, called gear, rack, or pinion, and a second body called the carrier (on which the gear is mounted), connected by a joint. One way to ensure that the geometry in your gear train maintains the desired spatial orientation during an analysis is to use the same body as the carrier body for gears to be mounted on. This is usually the ground or can be another body in the mechanism. Figure 6-1 shows a simple standard gear train in which the two parts used for the carrier bodies (two shafts) are mounted on the same body (in this case, the gear box housing that is not shown).

As shown in Figure 6-1, *Mechanism* simply uses cylinders or disks to represent the gears. No detailed tooth profile is necessary for any of the computations involved. Apparently, force and moment between a pair of teeth in contact will not be calculated in gear train simulations. However, there are other important data being calculated by *Mechanism*, such as reaction force exerting on the driven shaft (from a dynamic analysis), which is critical for mechanism design and analysis. Pitch circle diameters are important for defining gear trains in *Mechanism*.

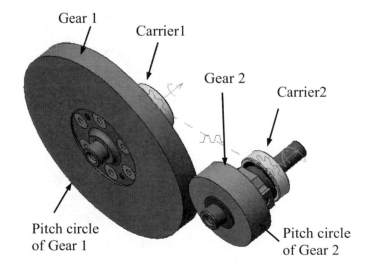

Figure 6-1 Gears and Carriers

Although cylinders or disks are sufficient to represent gears, we will use a more realistic gear train throughout this lesson. All gears in the example are shown with detailed geometric representation, including teeth. In addition, detailed parts, including shafts, bearing, screws and aligning pins are included for a realistic gear train system. In this gear train simulation, we will focus more on graphical animation, less on computations of physical quantities. We will add a servo motor to drive the input shaft, therefore conducting a kinematic analysis.

6.2 The Gear Train Example

Physical Model

The gear train example we are using for this lesson is part of a gearbox designed for an experimental lunar rover. The gear train is located in a gear box which is part of the transmission system of the rover, driven by a motor powered by solar energy. The purpose of the gear train is to convert high speed and small torque generated by the motor to low speed and large torque output in order to drive the wheels of the rover. The gear train consists of four spur gears mounted on three parallel axes, as shown in Figure 6-2.

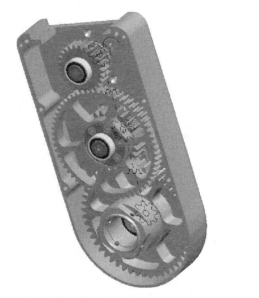

Figure 6-2 The Gear Train System in Rover

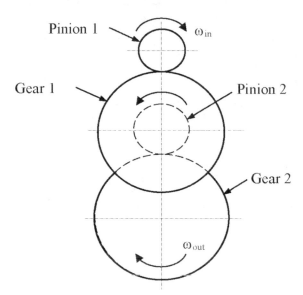

Figure 6-3 Schematic View of the Gear Train

The four spur gears form two gear pairs: *Pinion 1* and *Gear 1*, and *Pinion 2* and *Gear 2*, as illustrated in Figure 6-3. Note that *Pinion 1* is the driving gear that connects to the motor. The motor rotates in a clockwise direction, therefore driving *Pinion 1*. *Gear 1* is the driven gear of the first gear pair, which is mounted on the same shaft as *Pinion 2*. Both rotate in a counterclockwise direction. *Gear 2* is driven by *Pinion 2* and rotates in a clockwise direction. Note that the diameters of the pitch circles of the four gears are 50, 120, 60, and 125 mm, respectively; and the numbers of teeth are 25, 60, 24, and 50, respectively. The circular pitch P_c and module m of the two gear pairs are, respectively,

$$P_c = \frac{\pi d_{p1}}{N_{p1}} = \frac{\pi d_{g1}}{N_{g1}} = \frac{\pi(50)}{25} = \frac{\pi(120)}{60} = 6.283 \text{ mm}, \ m = \frac{d_{p1}}{N_{p1}} = \frac{d_{g1}}{N_{g1}} = \frac{50}{25} = \frac{120}{60} = 2 \text{ mm} \qquad (6.1a)$$

$$P_c = \frac{\pi d_{p2}}{N_{p2}} = \frac{\pi d_{g2}}{N_{g2}} = \frac{\pi(60)}{24} = \frac{\pi(125)}{50} = 7.854 \text{ mm}, \ m = \frac{d_{p1}}{N_{p1}} = \frac{d_{g1}}{N_{g1}} = \frac{50}{24} = \frac{125}{50} = 2.5 \text{ mm} \qquad (6.1b)$$

The velocity ratio Z of the gear train is:

$$Z = \frac{\omega_{out}}{\omega_{in}} = \frac{d_{p1}}{d_{g1}}\frac{d_{p2}}{d_{g2}} = \frac{50}{120}\frac{60}{125} = \frac{1}{5} \tag{6.2}$$

where ω_{out} and ω_{in} are the output and input angular velocities of the gear train system, respectively; and d_{p1}, d_{g1}, d_{p2}, and d_{g2} are the pitch circle diameters of the respective four gears. The velocity ratio of the gear train is 1:5; i.e., the angular velocity is reduced 5 times at the output. Theoretically, the torque output will increase 5 times if there is no loss due to, e.g., friction. Note that we will use *mmNs* unit system for this lesson.

Creo Parts and Assembly

The gear train assembly consists of one part and three subassemblies. You may want to open the final assembly, *gear_train_final.asm*, to check the assembled gear train shown in Figure 6-2. Enter *Mechanism* (by choosing *Applications* tab and clicking the *Mechanism* button 🔧), and choose *Playback* ◀▶ to show the gear motion. You should see that the pinion gear *Pinion 1* rotates a complete cycle, and the other three gears rotate accordingly. This is what we want to accomplish in this lesson.

Take a look at the model tree; there are four major components in this assembly: *gbox_housing.prt*, *gear_input.asm*, *gear_middle.asm*, and *gear_output.asm*. There are 22 distinct parts in this assembly.

One important thing for the animation to "look right" is to mesh the gear teeth properly. You may want to use the *FRONT* view and zoom in to the tooth mesh areas to check if the two pairs of gears mesh well (see Figure 6-4). Note that this is accomplished by properly orienting the three subassemblies with respect to the root assembly through datum plane alignment. Note that the tooth profile is represented by straight lines, instead of more accurate ones such as involute curves, for simplicity. Note that you may hide the gear housing part and rotate the view to see that the gear teeth meshed well between *Pinion 2* and *Gear 2*.

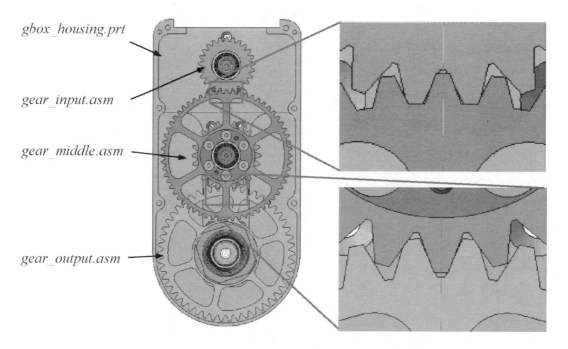

gbox_housing.prt

gear_input.asm

gear_middle.asm

gear_output.asm

Figure 6-4 Gear Teeth Properly Meshed

Simulation Model

In *Mechanism*, gear pairs are created by connecting two gears (or rack or pinion) and their carrier, as discussed earlier (see Figure 6-1). A pin joint must be created first between the gear and its carrier when you assemble the gear part. In this example, the gear housing part serves as the carrier for all gears. There are three pin joints to be created for the input, middle, and output gear assemblies. A motor will be added to drive the pin joint between the input gear subassembly and its carrier (the housing part). We will conduct a kinematic analysis for this example.

6.3 Using *Mechanism*

Creating an Assembly

In the *lesson6* folder, you should see a total of 22 distinct parts and 3 assemblies. The assemblies and associated parts are listed in Table 6-1.

Start *Creo*, select the working directory, and create a new assembly: *gear_train* (or a different name you prefer). You should see three datum planes and one datum coordinate system in the *Graphics* window. The first thing to do in this new assembly is to set unit system. From the *File* pull-down menu, choose *Prepare* > *Model Properties.*

Table 6-1 List of Parts and Assemblies in Lesson 6 Folder

Part/Subassemblies	Part Names	Remarks
gbox_housing.prt		
gear_input.asm	*wheel_gbox_shaft_input.prt*	
	wheel_gbox_pinion_1s.prt	*Pinion 1*
	spacer_12×18×5mm.prt	
	spacer_12×20×1mm.prt	
	bearing_12×18×8mm.prt (2)	
	spacer_10×18×014mm.prt	
	wheel_gbox_sft_mid_washer.prt	
	screw_tapper_head_5×15.prt	
	screw_set_tip_6×6.prt (2)	
gear_middle.asm	*wheel_gbox_pinion_2s.prt*	*Pinion 2*
	wheel_gbox_gear_1s.prt	*Gear 1*
	wheel_gbox_shaft_mid_pinion.prt	
	wheel_gbox_shaft_mid_gear.prt	
	bearing_12×18×8mm.prt (2)	
	screw_tapper_head_5×28.prt (6)	
	wheel_gbox_sft_mid_washer.prt (2)	
	screw_tapper_head_5×15.prt (2)	
	align_pin_4×27mm.prt (2)	
gear_output.asm	*wheel_gbox_gear_2s.prt*	*Gear 2*
	wheel_gbox_connect_wheel.prt	
	bear_tap_roller25×47×15mm.prt	
	screw_straight_head_4×15.prt (10)	
	align_pin_4×20mm.prt (2)	
	wheel_gbox_connect_wh_setscrew.prt (4)	

In the *Model Properties* dialog box, *Inch lbm Second (Creo Parametric Default)* is listed, as shown in Figure 6-5. Click *change* to bring up the *Units Manager* dialog box (Figure 6-6).

Figure 6-5 The *Model Properties* Dialog Box

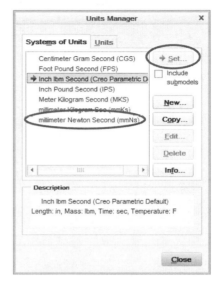

Figure 6-6 The *Units Manager* Dialog Box

In the *Units Manager* dialog box, choose *millimeter Newton second (mmNs)*, then click the *Set* button.

In the *Changing Model Units* dialog box, click *Interpret dimension (for example 1" becomes 1mm)*. In fact, it does not matter which option you choose since no component has been brought in yet. Click *OK* to accept the option (Figure 6-7) and click *Close* in the *Units Manager* dialog box to accept the unit system (Figure 6-6). Click *Close* again for the *Model Properties* dialog box (Figure 6-5).

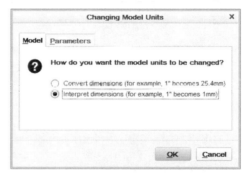

Figure 6-7 The *Changing Model Units* Dialog Box

Now we are ready to bring in the first part *gear_housing.prt*. The gear housing will be fixed to the assembly by aligning their respective coordinate systems.

Click the *Assemble* button ⬚ and choose *gear_housing.prt*. In the *Component Placement* dashboard choose *Coincident* from the *Constraint Type* list. From the *Graphics* window, pick *ASM_DEF_CSYS* (assembly) and *PRT_DEF_CSYS* (*gear_housing.prt*), as shown in Figure 6-8. Then click ✓ to accept the definition.

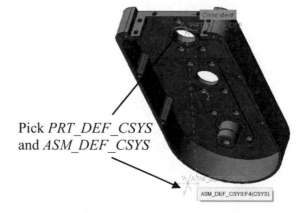

Pick *PRT_DEF_CSYS* and *ASM_DEF_CSYS*

Figure 6-8 Assemble the gear housing

Three datum axes of the housing, *A_191*, *A_189*, and *A_160* of the three features *Cut id 10858*, *Cut id 10788*, and *Protrusion id 10035*, respectively, will be used to align with axes of the respective gears for creating pin joints. There are many axes in this part. In order to see these three datum axes, you may want to use the layer option to filter out some of them.

Click the *View* tab, and choose the *Layers* button ⬛ to display the *Layer Tree* in the navigation, or from *Model Tree* navigator window click *Show* button ⬛ and then choose *Layer Tree* (see Figure 6-9). As shown in Figure 6-10, the *Model Tree* area will now display default layers created by *Creo*. Expand *02__PRT_ALL_AXES*, and then click *in GBOX_HOUSING.PRT*; a number of solid features that contain datum axes will be listed.

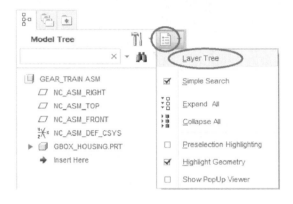

Figure 6-9 The *Show* Button and
Layer Tree Option

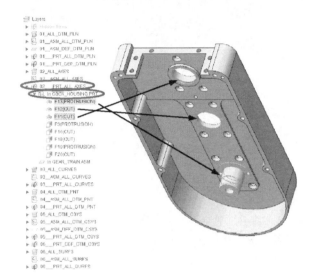

Figure 6-10 Solid Features *F13(PROTRUSION)*,
F12(CUT), and *F11(CUT)*

Click the solid features *F13(PROTRUSION)*, *F12(CUT)*, and *F11(CUT)* to locate them in the *Graphics* window. We will use the axes embedded in these three solid features.

Use the right mouse button to hide all the other items. You should see a cleaner view of the datum axes with the part. Click *Show* button ⬛ and choose *Model Tree* to switch back to the standard *Model Tree*.

Next, we will bring in the first gear assembly, *gbox_input.asm*. Click the *Assemble* button ⬛ and choose *gbox_input.asm*. In the *Component Placement* dashboard from the *User Defined* list choose *Pin*. Pick *A_191* (housing) and *A_21* (*gbox_input.asm*), as shown in Figure 6-11a. Now turn off the datum axis display. Rotate the view (similar to Figure 6-11b) to pick the back face of the housing and the groove face of the *gbox_input.asm* (*Bearing_12×28×8MM: Surf:F5(PRUTRUSION)*), as shown in Figure 6-11b. The *gbox_input.asm* should be properly assembled to the housing with a pin joint, as shown in Figure 6-11c. Click ✔ to accept the definition.

The next component we will bring in is *gbox_middle.asm*. Click the *Assemble* button and choose *gbox_middle.asm*. Reorient the *gbox_middle.asm* (similar to that of Figure 6-12a). From the *User Defined* list choose *Pin*. Turn on datum axis display, and pick *A_189* (*gear_housing.prt*) and *A_66* of *WHEEL_GBOX_SFT_MID_WASHER* in *gbox_middle.asm*, as shown in Figure 6-12a. Note that you may need to zoom in for a closer view and look for axis *WHEEL_GBOX_SFT_MID_WASHER:*

A 66(AXIS):F5(PROTRUSION). Now turn off the datum axis display, pick the back face of the housing and the groove face of *gbox_middle.asm* (*F5* of *Bearing 12×28×8*), as shown in Figure 6-12b. The *gbox_middle.asm* should be properly assembled to the housing with a pin joint, as shown in Figure 6-12c. Click ✅ to accept the definition.

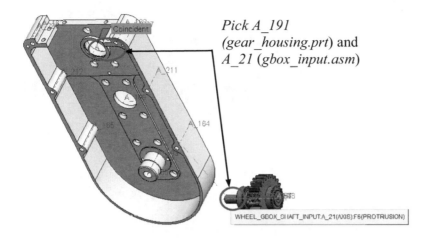

Pick A_191
(*gear_housing.prt*) and
A_21 (*gbox_input.asm*)

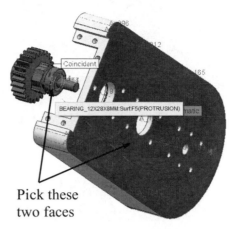

Pick these
two faces

(a) Pick the Two Datum Axes

(b) Pick the Two Faces

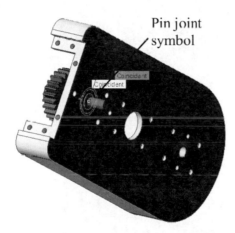

Pin joint
symbol

(c) *gbox_input.asm* Assembled

Figure 6-11 Assembling *gbox_input.asm* to Gear Housing

The third and final assembly we will bring in is *gbox_output.asm*. Following the same procedure to choose a pin joint type, pick *A_193* (*gear_housing.prt*) and *A_66* of *gbox_output.asm* (*WHEEL_GBOX_CONNECT_WHEEL:A_66(AXIS):F5 (PROTRUSION)*), as shown in Figure 6-13a. Now turn off the datum axis display, pick the back face of *gbox_output.asm* (*F71* of *wheel_gbox_gear_25*), as shown in Figure 6-13b. Then, rotate the view to pick the inner face of the housing, as shown in Figure 6-13b. The *gbox_output.asm* is assembled to the housing. If *gbox_output.asm* is not properly assembled (like that of Figure 6-13c), click the *Change orientation of constraint* button ⤢ on top of the *Graphics* window (circled in Figure 6-13c) to flip its orientation. The *gbox_output.asm* should now be properly assembled with a pin joint, as shown in Figure 6-13d. Note that you may change the constraint type from *Coincident* to *Distance* and enter 5 mm to align the output gear (*Gear 2*) with *Pinion 2* of the middle gear assembly more accurately. Click ✅ to accept the definition.

Pick *A_189*
(*gear_housing.prt*) and
A_66 (*gbox_middle.asm*)

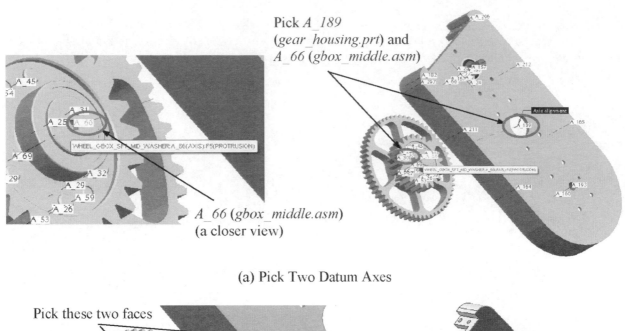

A_66 (*gbox_middle.asm*)
(a closer view)

(a) Pick Two Datum Axes

Pick these two faces

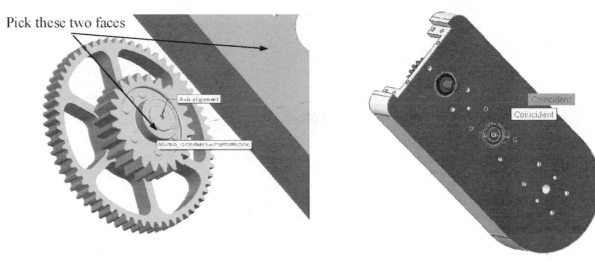

(b) Pick Two Faces   (c) *gbox_middle.asm* Assembled

Figure 6-12 Assembling *gbox_middle.asm* to Housing

Now we have completed the assembly of the gear train. You may want to hide the housing and rotate the view to see more closely how these gears are assembled (Figure 6-14). Note that the gear teeth between *Pinion 1* and *Gear 1* are meshed well, but not those between *Pinion 2* and *Gear 2*. We will use the *Drag* option to align datum planes of the gears so that their teeth would mesh properly. Before that we will have to hide some of the datum planes and only keep those we need to use for aligning the gear teeth.

First turn on datum planes display; you will see a huge number of datum planes appear. Now go to the *Model Tree* window. Expand all three subassemblies, and hide all parts, except for *wheel_gbox_pinion_1s.prt* (in *gbox_input.asm*), *wheel_gbox_gear_1s.prt* (in *gbox_middle.asm*), and *wheel_gbox_gear_2s.prt* (in *gbox_output.asm*). The best way to do that is to select all parts you want to hide (press *Shift* or *Ctrl* key to select multiple items), press the right mouse key, and click the *Hide* button , as shown in Figure 6-15.

Follow same approach to hide datum planes of the three subassemblies. In order to display datum features in the model tree, you may click the *Settings* button , and then click *Tree Filters* button (on top of the model tree shown in Figure 6-16). In the *Model Tree Filters* dialog box (Figure 6-17), click *Features* and then *OK*. All features, including datum planes, will now appear in the model tree.

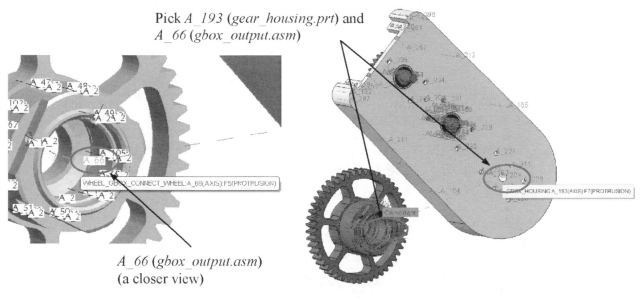

Pick *A_193* (*gear_housing.prt*) and *A_66* (*gbox_output.asm*)

A_66 (*gbox_output.asm*) (a closer view)

(a) Pick Two Datum Axes

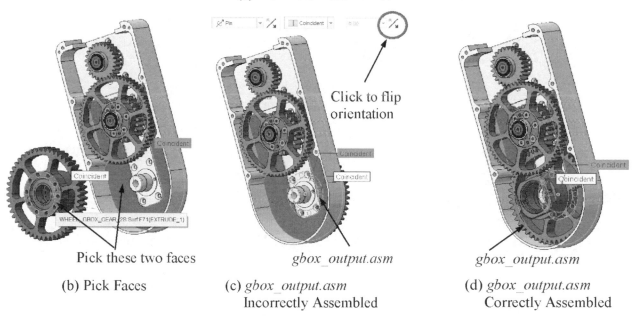

Click to flip orientation

Pick these two faces

gbox_output.asm

gbox_output.asm

(b) Pick Faces

(c) *gbox_output.asm* Incorrectly Assembled

(d) *gbox_output.asm* Correctly Assembled

Figure 6-13 Assembling *gbox_output.asm* to Housing

Also, hide *ASM_TOP* and *ASM_FRONT* of the root assembly. You should see a cleaner view like that of Figure 6-18a.

Click the *Drag Components* button  on top of the *Graphics* window; the *Drag* dialog box appears. Click the *Constraints* tab and choose the *Align Two Entities* button (first on the left). Choose datum planes *SIDE* (*wheel_gbox_pinion_1s*) of *gbox_input.asm* and *ASM_RIGHT* (assembly), as shown in Figure 6-18a. Click the *Align Two Entities* button again, and choose *SIDE* (*wheel_gbox_gear_1s*) of *gbox_middle.asm* and *ASM_RIGHT* (assembly), as shown in Figure 6-18b. Click the *Align Two Entities* button again, and choose *TOP* (*wheel_gbox_gear_2s*) of *gbox_output.asm* and *ASM_RIGHT* (assembly), as shown in Figure 6-18c.

Click the *Current Snapshot* button on top; a default name *Snapshot1* will appear. Click *Close*. The snapshot *Snapshot1* has been saved for future use.

Unhide *wheel_gbox_pinion_2s* of the *gear_middle.asm* by pressing the right mouse button and choosing *Unhide*. Change to the *FRONT* view and zoom in to check the tooth meshing areas. All teeth should now be properly meshed.

Now the gear train is completely assembled. Save your model before we enter *Mechanism*.

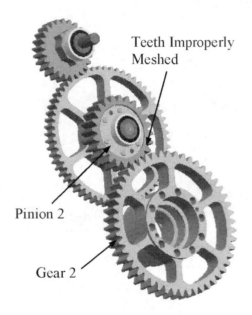

Figure 6-14 The Gear Train

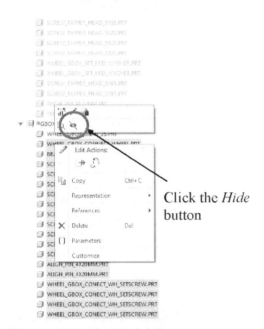

Figure 6-15 The Model Tree

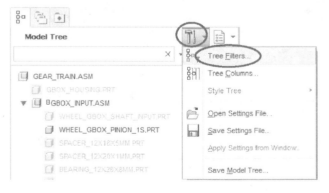

Figure 6-16 Choose the Tree Filters

Figure 6-17 The *Model Tree Filters* dialog box

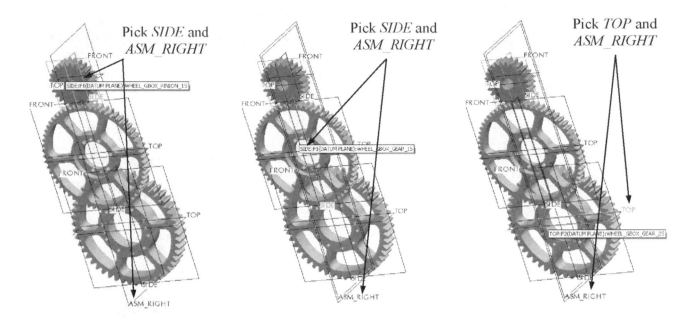

(a) Align *SIDE* and *ASM_RIGHT*   (b) Align *SIDE* and *ASM_RIGHT*   (c) Align *TOP* and *ASM_RIGHT*

Figure 6-18 Aligning Datum Planes for Meshing Gear Teeth

Creating a Simulation Model

We are now ready to enter *Mechanism* and create a dynamic simulation model. Enter *Mechanism* by clicking the *Applications* tab and choosing *Mechanism* button .Turn off all datum feature display; you should see three pin joints shown in the *Graphics* window (see Figure 6-19). Note the third pin joint is in the opposite direction of the first two.

We will define two gear pairs and a servo motor for a motion model.

Click the *Gears* button of the *Connections* group at the top of the *Graphics* window. The *Gear Pair Definition* dialog box will appear (Figure 6-20a). Use the default name (*GearPair1*), choose *Spur* for *Type*, then pick the first pin joint *Pin1*. Note that you may want to turn off all datum features in order to see the pin joint symbols. After picking the pin joint, *gbox_input.asm* and *gbox_housing.prt* will be highlighted. Enter *50* for *Pitch Circle Diameter*. Note the direction arrow (double arrowhead in pink color) of Gear 1 is consistent with that of the *Pin* joint.

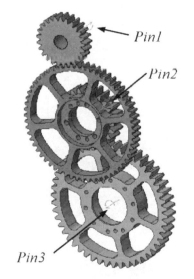

Figure 6-19 Pin Joints Defined in the Assembly

Click the *Gear2* tab to define the second gear in the pair (see Figure 6-20b). Pick *Pin2* and enter *120* for *Pitch Circle Diameter*. Note the direction arrow of Gear 2 is reversed to that of Gear 1 (see Figure 6-21), which is correct since gears are externally meshed. If this is not the case, you will need to click the *Flip* button next to the joint axis field (see Figure 6-21) to flip its direction.

Click *OK*. A pair of gear symbols will appear next to the two pin joints chosen (*Pin1* and *Pin2*), as shown in Figure 6-22.

Repeat the same procedure to define the second pair. Pick *Pin2* and *Pin3* and enter diameters *60* and *125* for *Gear1* and *Gear2*, respectively.

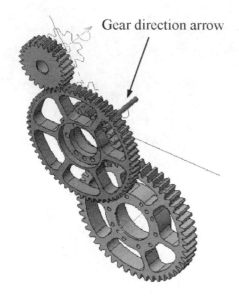

Figure 6-21 Gear Direction Arrow

(a) *Gear1* Tab   (b) *Gear2* Tab

Figure 6-20 The *Gear Pair Definition* Dialog Box

Figure 6-22 The *Gear Pair* Symbols

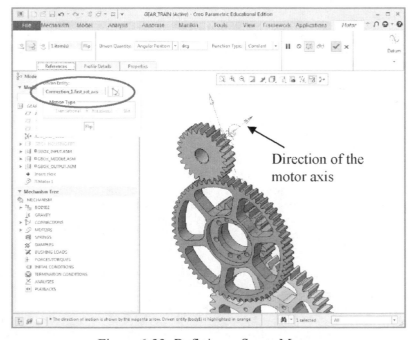

Figure 6-23 Defining a Servo Motor

Click the *Drag* button 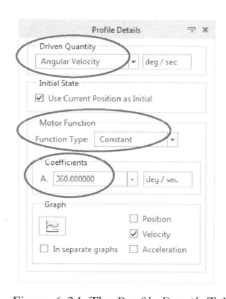 and click *Pinion 1*. Move the mouse around; you should see that *Pinion 1* starts turning, and all gears in the gear train move as expected.

Now we will create a driver at pin joint *Pin1* to drive *Pinion 1*, therefore, the gear train.

Click the *Servo Motors* button of the *Insert* group on top of the *Graphics* window; a new set of selections will appear (see Figure 6-23). Pick *Pin1* from the *Graphics* window. After picking the pin joint, a larger arrow appears to confirm your selection. The rotation axis must point in a direction like that of Figure 6-23. You may need to click the *Flip* button to make sure you pick the correct direction of the rotation axis. Click the *References* tab to make sure that *Connection_1.first_rot_axis* has been selected for *Driven Entity*, as circled in Figure 6-23.

The next step is to specify the profile of the motor. Click the *Profile Details* tab (next to the *References* tab), choose *Angular Velocity* for *Driven Quantity* and *Constant* (default) for *Function Type* (see Figure 6-24). Enter *360* for the constant *A* under *Coefficients*.

Click the *Properties* tab to enter *Motor1* for name. Click the ✔ button on the right to accept the definition. A motor symbol appears in the *Graphics* window overlapping with *Pin1*, as circled in Figure 6-25.

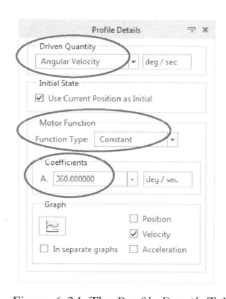

Figure 6-24 The *Profile Details* Tab

Servo Motor (or Driver)

Figure 6-25 *Servo Motor* Symbol

Creating and Running a Kinematic Analysis

Click the *Mechanism Analysis* button to define an analysis. In the *Analysis Definition* dialog box appearing, leave the default name, *AnalysisDefinition1*, and select or enter:

Type: *Kinematic*
Start Time: *0*
End Time: *1*
Frame Rate: *100*
Minimum Interval: *0.01*

Initial Configuration: *Current*

Make sure the servo motor is included in the analysis. Use the saved snapshot (*Snapshot1*) for initial configuration before defining and running the analysis.

Run the analysis. In the *Graphics* window, the mechanism starts moving. You should see *Pinion 1* rotates 360 degrees as expected. Click *OK* to close the dialog box.

Saving and Reviewing Results

Click the *Playback* ◀▶ of the *Analysis* group to bring up the *Playbacks* dialog box and repeat the motion animation. On the *Playbacks* dialog box, click the *Save* button 💾 to save the results as a *.pbk* file.

Note that we want to create a measure to monitor the output angular velocity of the gear train. We will choose the pin joint *Pin3* and the angular velocity of its rotational axis for the measure. To do so, click the *Measures* button ⊠ of the *Analysis* group. In the *Measure Results* dialog box appearing (Figure 6-26), click the *Create New Measure* button 🗋. The *Measure Definition* dialog box opens. Enter *Angular_Velocity_Output* for *Name*. Under *Type*, select *Velocity*. Pick *Pin3* in the *Graphics* window. Under *Evaluation Method*, leave *Each Time Step*. Click *OK* to accept the definition.

In the *Measure Results* dialog box, choose *AnalysisDefinition1* under the *Result Set* and click the *Graph* button ⊠ on the top left corner to graph the measure. The graph should be similar to that of Figure 6-27, which shows that the output velocity is a constant of −72 degrees/sec. Note that this magnitude is one fifth of the input velocity since the gear ratio is 1:5. The negative sign is simply due to the direction of the joint *Pin3*. Physically, both the input (*Pinion 1*) and output gears (*Gear 2*) rotate in the same direction.

Figure 6-26 The *Measure Definition*
Dialog Box

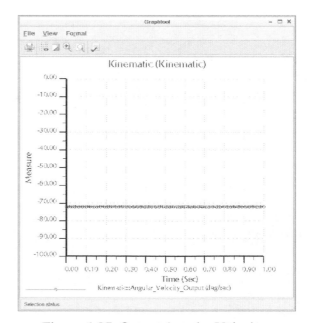

Figure 6-27 Output Angular Velocity

Exercises:

1.   The same spur gear train will be used for this exercise. Create a constant torque for the input gear (*gbox_input.asm*) along the *Z*-direction of the *WCS*. Define and run a 2-second dynamic simulation for the gear train.

   (a)   What is the minimum torque that is required to rotate the input gear, and therefore, the entire gear train?

   (b)   If the torque applied to the input gear is 100 mm N, what will be the maximum output angular velocity of the gear train at the end of the 2-second simulation? Verify the simulation result using your own calculation. Note that you may check mass properties of the bodies from *Model Tree*. Click the body name and use right mouse button to choose *Info > Details*.

   (c)   Create a graph for the reaction moment between gears of the first gear pair due to the 100 mm N torque. What is the reaction moment obtained from simulation?

Notes:

Lesson 7: Planetary Gear Train Systems

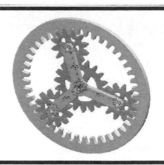

7.1 Overview of the Lesson

In this lesson we will discuss planetary gear train systems. Planetary gearing is a gear system that consists of one or more outer gears, or *planet* gears, rotating about a central, or *sun* gear, as shown in Figure 7-1. Typically, the planet gears are mounted on a movable arm (or carrier) which may rotate relatively to the sun gear. A planetary gear system also incorporates the use of an outer ring gear (or annulus), which meshes with the planet gears.

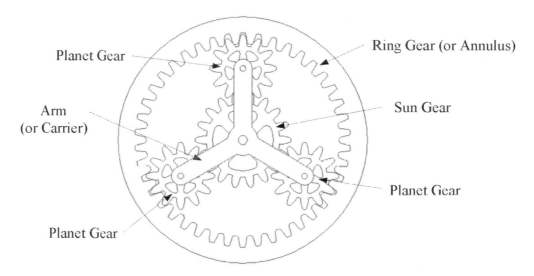

Figure 7-1 Planetary Gear Train

In many planetary gearing systems, one of these three basic components (that is, sun gear, planet gears, and arm) is held stationary (in addition to the ring gear, which is usually fixed); one of the two remaining components is an *input*, receiving power to the gear train system, while the last component is an *output*, disseminating power from the system. The ratio of input rotation to output rotation is dependent upon the number of teeth in meshed gear pairs, and upon which the component is held stationary. One situation is when the ring gear is held stationary, and the sun gear is used as input. In this case, the planet gears rotate around the sun gear at a rate determined by the number of teeth in meshed gear pairs, i.e., sun/planet and planet/ring gears. The same is true for the angular velocity of the arm.

Planetary gear trains are commonly employed in mechanical systems. They are primarily designed for large gear reductions, hence building up sufficient torque to drive the mechanism system. One of the common places to find the planetary gear trains is in the power screwdriver (Figure 7-2), where a large gear reduction is necessary for a large torque output in a small space. Series of planetary gear trains are often packaged to receive a maximum gear reduction.

In general, one of the most important tasks in designing planetary gear train systems is to achieve a required gear reduction, usually in a compact space. In this lesson, we will learn how to simulate motion of planetary gear trains and calculate the velocity ratios. Note that some calculation results obtained from *Mechanism* for the planetary gear trains are not reliable. They are wrong in fact. Therefore, we will discuss the theory to make correct calculations. It is very important that you understand the theory and are able to compute the velocity ratio of the gear train correctly.

7.2 The Planetary Gear Train Examples

Physical Model

Figure 7-2 Planetary Gear Trains Found in Power Screw Driver

We discuss two examples. The first example is a single planetary gear train, as shown in Figure 7-3. This gear train consists of only one planet gear mounted on the arm. The planet gear is meshed with a stationary ring gear. In this system, there is only one pair of meshed gears: the planet and the ring gears. The pitch circle diameters of the planet and ring gears are $d_p = 2.333$ and $d_r = 8.167$ in., respectively. There are 12 and 42 teeth on the planet and ring gears, respectively. Therefore, the circular pitch P_c and diametral pitch P_d of the gears are, respectively,

$$p_c = \frac{\pi d_p}{N_p} = \frac{\pi d_r}{N_r} = \frac{\pi(2.333)}{12} = \frac{\pi(8.167)}{42} = 0.6108 \text{ in.} \qquad (7.1a)$$

$$p_d = \frac{N_p}{d_p} = \frac{N_r}{d_r} = \frac{12}{2.333} = \frac{42}{8.167} = 5.143 \qquad (7.1b)$$

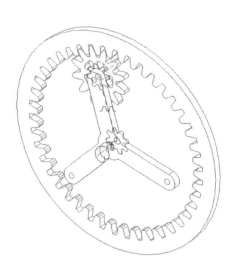

Figure 7-3 The Single Planetary
Gear Train System

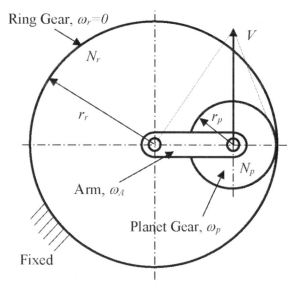

Figure 7-4 Velocity Analysis for the Single
Planetary Gear Train System

Note that every one counterclockwise turn of the arm produces $(1-N_r/N_p)$ clockwise turns of the planet gear. This can be easily figured out from velocity analysis. For example, as shown in Figure 7-4, the linear velocity at the end of the arm where it connects to the center of the planet gear is $V = (r_r - r_p)\omega_A$.

Since the planet gear is "rolling" along the ring gear, the linear velocity at its center, which is the same as V, is $V = r_p (-\omega_p)$. Note that the negative sign is added to match the signs of the angular velocities. Therefore,

$$V = \left(r_r - r_p\right)\omega_A = r_p\left(-\omega_p\right) \tag{7.2a}$$

$$\omega_p = \left(1 - \frac{r_r}{r_p}\right)\omega_A = \left(1 - \frac{N_r}{N_p}\right)\omega_A \tag{7.2b}$$

The velocity ratio Z of the gear train is:

$$Z = \frac{\omega_{out}}{\omega_{in}} = \frac{\omega_p}{\omega_A} = 1 - \frac{N_r}{N_p} = 1 - \frac{42}{12} = -2.5 \tag{7.3}$$

That is, every one counterclockwise turn of the arm produces 2.5 clockwise turns on the planet gear. If the arm is rotating at 360 degrees/sec counterclockwise, the planet gear will rotate at -900 (that is, -2.5×360) degrees/sec (clockwise).

The second example, as shown in Figure 7.5, consists of an additional sun gear and two additional planet gears. Again, the ring gear is stationary and the sun gear is driven by a motor of constant angular velocity. The pitch circle diameter of the sun gear is $d_s = 3.5$ and the number of teeth is $N_s = 18$. Therefore, the circular pitch P_c and diametral pitch P_d of the sun gear are 0.6108 and 5.143 in., respectively, which are identical to the remaining gears in the system, as expected.

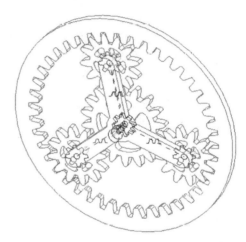

Note that for a general planetary gear train system, the arm is usually not directly driven by the power source (such as a motor). The general equation for calculating the gear ratio is the following:

$$Z = \frac{\omega_{out} - \omega_A}{\omega_{in} - \omega_A} = \prod\left(\pm\frac{N_{driving}}{N_{driven}}\right) \tag{7.4}$$

Figure 7-5 The Multiple Planetary Gear Train System

where $N_{driving}$ and N_{driven} are the number of teeth on the driving and driven gears, respectively, of a gear pair; and $\prod$ is an operator that creates a product for the terms enclosed in the parentheses. Note that the driving and driven gears can be identified by power input and output within each gear pair. For example, for the gear system shown in Figure 7-5, if the sun gear is directly connected to a motor, then the sun gear will be the driving gear and the planet gear (any one of the three) is the driven gear. In the second gear pair, i.e., the planet and the ring gears, the planet gear will be the driving gear and the ring gear will be the driven gear. The sign on the right-hand side of Eq. 7.4 will be determined by how the gear pairs are meshed, externally or internally. For a gear pair consisting of two regular spur gears, such as the one discussed in *Lesson 6*, the sign is negative since the two spur gears are meshed externally and rotate in the opposite directions. For a gear pair including a ring gear, the sign will be positive since, in this case, the planet and the ring gears are meshed internally and rotate in the same direction. Therefore, for the second example, if the sun gear is driven by a motor with a constant angular velocity of 360 degrees/sec counterclockwise, what will be the gear ratio of the system? Well, we know the output gear, i.e., the ring gear, is stationary; therefore, $\omega_{out} = 0$. In order to calculate the gear ratio, we will have to figure out the angular velocity of the arm first. This can be accomplished by using Eq. 7.4.

$$\frac{\omega_{out}-\omega_A}{\omega_{in}-\omega_A} = \frac{0-\omega_A}{360-\omega_A} = \left(-\frac{N_s}{N_p}\right)\left(\frac{N_p}{N_r}\right) = \left(-\frac{18}{12}\right)\left(\frac{12}{42}\right) = -0.4286 \tag{7.5}$$

Therefore, $\omega_A = 108.0$ degree/sec counterclockwise, and the velocity ratio is -0.4286 if the input and output gears of the system are the sun and ring gears, respectively.

If the sun gear is the input gear and the planet gear is the output gear, then the angular velocity of the planet gear can be calculated by using Eq. 7.4; i.e.,

$$\frac{\omega_{out}-\omega_A}{\omega_{in}-\omega_A} = \frac{\omega_p-108}{360-108} = \left(-\frac{N_s}{N_p}\right) = \left(-\frac{18}{12}\right) = -1.5 \tag{7.6}$$

Therefore, $\omega_p = -270.0$ degrees/sec (clockwise), and the velocity ratio is -1.5.

Creo Parts and Assembly

The single planetary gear train system consists of three parts: *arm.prt*, *plant.prt*, and *ring.prt*. You may want to open the final assembly, *single_gear_train_final.asm*, to check the assembled gear train shown in Figure 7-3. Enter *Mechanism* (by choosing *Applications > Mechanism*), choose *Analysis > Playback* to show the gear motion. You should see that the arm rotates and drives the planet gear to rotate around the inner side of the ring gear. This is the first example we want to accomplish in this lesson.

The multiple planetary gear train consists of *sun.prt*, *arm.prt*, *plant.prt (3)*, and *ring.prt*. You may want to open the final assembly, *multiple_gear_train_final.asm*, to check the assembled gear train shown in Figure 7-5. Again, choose *Analysis > Playback* to show the gear motion.

Similar to what was discussed in *Lesson 6*, one important thing for the animation to "look right" is that the gear pairs must mesh correctly. You may want to use the *BACK* view and zoom in to the tooth mesh areas to check if the two pairs of gears mesh well (see Figure 7-6). They are properly meshed. This is accomplished by properly orienting the individual gear parts with respect to the assembly through datum plane alignment.

Note that the example files you downloaded from the publisher's web site should consist of two subfolders, *single* and *multiple*. The files enclosed in the folders are, respectively:

lesson7/final/single/single_gear_train_final.asm
lesson7/final/single/arm.prt
lesson7/final/single/planet.prt
lesson7/final/single/ring.prt

and

lesson7/final/multiple/multiple_gear_train_final.asm
lesson7/final/multiple/sun.prt
lesson7/final/multiple/arm.prt
lesson7/final/multiple/planet.prt
lesson7/final/multiple/ring.prt

Note that in this lesson, we will use the default unit system, i.e., in-lb$_m$-sec.

Figure 7-6 Gear Teeth Properly Meshed

Simulation Model

As discussed in *Lesson 6*, in *Mechanism*, gear pairs are created by selecting two gears and their carriers. A pin joint must be created first between the gear and its carrier when you assemble the gear. In both examples, the datum axis (*AA_1*, to be created in assembly) will be the carrier for both the sun and ring gears. In addition, *AA_1* will be chosen to define a pin joint for the arm. Only one gear pair will be defined for the single planetary gear train example, consisting of the planet and ring gears. There are multiple gear pairs defined for the multiple planetary gear train example, the sun and planet gears, and planet and ring gears.

Two motors will be created for both examples. One motor will be used to drive the arm and the sun gear for the single and multiple planetary gear train examples, respectively, at a constant angular velocity. The second motor will be connected to the ring gear, in which the angular velocity will be set to zero since the ring gear is assumed stationary. Additional motors will be created for the multiple planetary gear train example.

7.3 Using *Mechanism*

Creating an Assembly

Start *Creo*, select working directory, and create a new assembly: *Example1* (or any name you prefer). You should see three datum planes and one datum coordinate system in the *Graphics* window. We will first create a datum axis that is normal to the datum plane *ASM_FRONT*, and lies on both *ASM_TOP* and *ASM_RIGHT*.

To create a datum axis, click the *Axis* button ⬚ of the *Datum* group on top of the *Graphics* window. In the *Datum Axis* dialog box appearing (Figure 7-7), the *References* field is active and is ready for you to pick a reference entity from the *Graphics* window. Pick *ASM_FRONT*; a temporary axis that is normal to *ASM_FRONT* will appear with two handles (small half-squares), as shown in Figure 7-8. Drag one handle to touch *ASM_TOP*. When the *ASM_TOP* is highlighted, release the mouse button. In the *Graphics* window, you should see a dimension appears, defining the distance between the axis and *ASM_TOP*. In the *Datum Axis* dialog box, *ASM_TOP* is listed in the *Offset references* with an offset value. Enter *0* for the offset. Repeat the same to drag the other handle to lie on the datum plane *ASM_RIGHT*. Set the offset value to *0*, and click *OK*. You should see a datum axis *AA_1* created in the *Graphics* window. Note that datum axis *AA_1* will be the carrier for the ring gear, as well as the rotation axis of the pin joint between the arm and the assembly.

Figure 7-7 The *Datum Axis*
Dialog Box

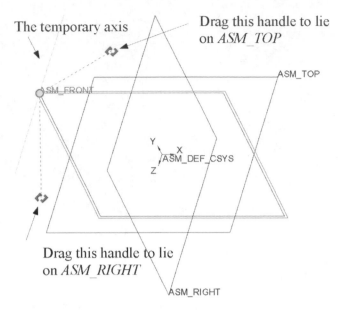

Figure 7-8 Creating a *Datum Axis*

Next, we will bring in the first part, *arm.prt*. Click the *Assemble* button 🔳 and choose *arm.prt*. In the *Component Placement* dashboard from the *User Defined* list, choose the *Pin* joint, and pick *AA_1* (assembly) and *A_3* (*arm.prt*), as shown in Figure 7-9a. Now turn off datum axis display and turn on datum plane display. Pick two datum planes, *ASM_FRONT* (assembly) and *FRONT* (*arm.prt*), as shown in Figure 7-9b. The *arm.prt* should be properly assembled to the assembly through a pin joint. The default name of the pin joint is *Connection_1*. In order to minimize confusion (*Mechanism* may give the same name to two different joints), we will rename this joint *Pin1*. This can be done by clicking the *Placement* button, clicking the joint name *Connection_1*, and entering *Pin1* in the *Set Name* text field, as shown in Figure 7-10. Press the *Enter* key to accept the joint name. Click ✅ to accept the joint definition.

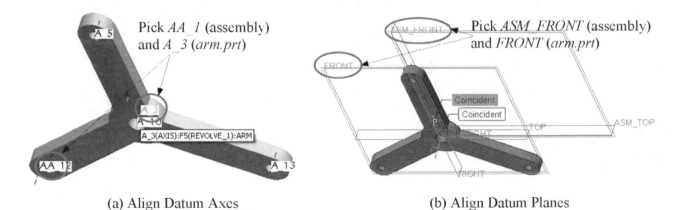

(a) Align Datum Axes   (b) Align Datum Planes

Figure 7-9 Assembling *arm.prt*

The next part we will bring in is the planet gear. Click the *Assemble* button 🔳 and choose *planet.prt*. Choose the *Pin* joint, turn on datum axis display, and pick *A_2* (*planet.prt*) and *A_5* (*arm.prt*), as shown in Figure 7-11a. Pick the front face of the planet gear, *F5(REVOLVE_1):PLANET*, and rotate the view to pick the back face of the arm, *F5(REVOLVE_1):ARM*, as shown in Figure 7-11b. The *planet.prt* should be properly assembled to the arm with a pin joint. Change the joint name to *Pin2*. Click ✔ to accept the definition.

Figure 7-10 Changing the Joint Name to *Pin1*

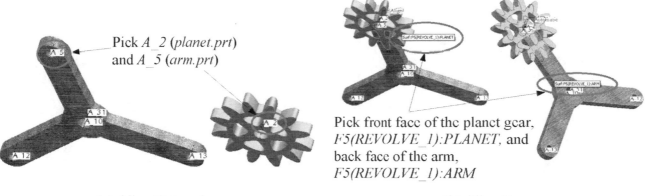

(a) Align Datum Axes   (b) Align Faces

Figure 7-11 Assembling *planet.prt*

The third and final part we will bring in is *ring.prt*. Following the same procedure to choose a pin joint type, pick *AA_1* (assembly) and *A_3* (*ring.prt*), as shown in Figure 7-12a. Now turn off the datum axis display, turn on datum plane display, and pick *ASM_FRONT* (assembly) and *FRONT* (*ring.prt*), as shown in Figure 7-12b. The ring is not properly aligned with the planet gear yet. Click the *Placement* button, select *Translation* (should be selected already) and select *Distance* (see Figure 7-12c), and enter –0.3125 for *Offset*. Note that this value is the thickness of the arm. With this offset, the ring gear will be properly aligned with the planet gear. Enter *Pin3* for name. The *ring.prt* should now be properly assembled to the assembly with a pin joint. Again, click ✔ to accept the definition.

The assembly is now complete. You may want to rotate the view to see if these gears are meshed properly. Note that the gear teeth between planet and ring gears should mesh well. We will capture this configuration, save it as a snapshot, and use the snapshot as the initial condition for motion analysis. You may use the *Drag Components* button 🖲 at the top of the *Graphics* window to save the current configuration as a snapshot. In the *Drag* dialog box appearing, click the *Current Snapshots* button 📷 at the top; a default name *Snapshot1* will appear. Click *Close*. The snapshot *Snapshot1* is saved for future use. Now we are ready to carry out a motion simulation for the gear train system. Save your model before we move into *Mechanism*.

Pick *AA_1* (*assembly*)
and *A_3* (*ring.prt*)

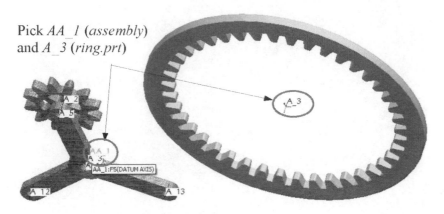

(a) Align Datum Axes

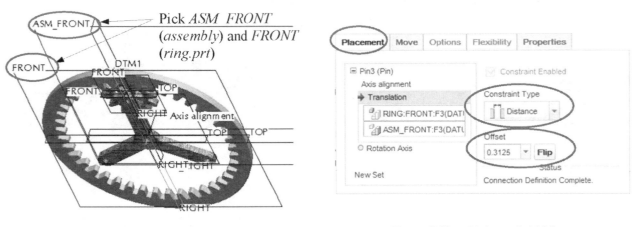

Pick *ASM_FRONT*
(*assembly*) and *FRONT*
(*ring.prt*)

(b) Align Datum Planes   (c) Enter Offset Value –*0.3125*

Figure 7-12 Assembling *ring.prt*

Creating a Motion Model

We are now ready to enter *Mechanism* for creating a dynamic simulation model. To enter *Mechanism*, simply click the *Applications* tab on top of the *Graphics* window, and choose *Mechanism* button.

There are three pin joints that appear, as shown in Figure 7-13, including two pin joints, *Pin1* and *Pin3*, overlapping at the center of the gear train system. Click the *CONNECTIONS* entity in the *Mechanism* model tree (lower half) to expand its contents, and then click *JOINTS*. There should be three connections listed: *Pin1*, *Pin2*, and *Pin3* (Figure 7-14a). Click *Pin1* and then *Pin3* to see the joint symbols highlighted in the *Graphics* window. Expand joint *Pin1*; you should see that *Pin1* is defined between *Ground* and *body1* (that is, *arm.prt*), and so on. Since there are two joints overlapping, we have to be very careful in picking the correct joints for defining the motion model. Also, if we expand the *BODIES* entities, there are *Ground*, *body1*, *body2*, and *body3* listed, as shown in Figure 7-14b. Note that *body1* is the arm, *body2* is the planet gear, and *body3* is the ring gear.

We will define one gear pair and two motors for a motion model. The gear pair consists of the planet and ring gears. The motor defined at joint *Pin1* will have a constant angular velocity driving the arm. The second motor will be defined at *Pin3* and is stationary (since the ring gear is assumed stationary).

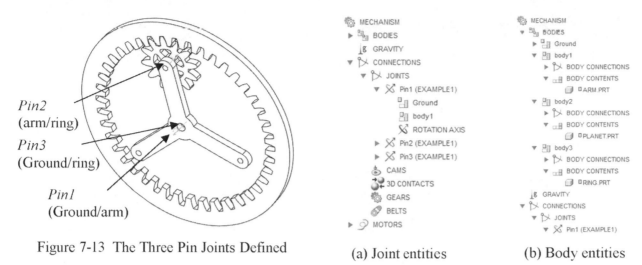

Pin2
(arm/ring)

Pin3
(Ground/ring)

Pin1
(Ground/arm)

Figure 7-13 The Three Pin Joints Defined

(a) Joint entities   (b) Body entities

Figure 7-14 Motion Entities in the Model Tree

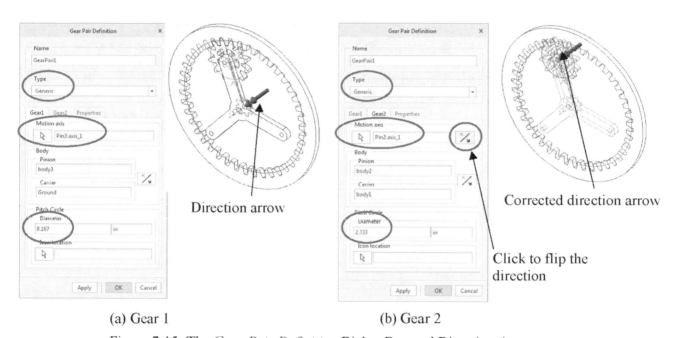

Direction arrow

Corrected direction arrow

Click to flip the
direction

(a) Gear 1   (b) Gear 2

Figure 7-15 The *Gear Pair Definition* Dialog Box and Direction Arrows

Click the *Gears* button of the *Connections* group at top of the *Graphics* window. The *Gear Pair Definition* dialog box will appear (Figure 7-15). Use the default name (*GearPair1*), leave *Generic* (default) for *Type*, and then pick the joint *Pin3* (between the ring gear and the ground) from the *Graphics* window. Note that you may want to turn off all datum features in order to see the pin joint. If you picked a wrong joint, simply click the arrow button under the *Motion axis* in the *Gear Pair Definition* dialog box (Figure 7-15a), and pick again. The best way to pick the correct joint is to move the mouse pointer close to the joints, and wait for a few seconds until a label showing the joint name appears. If the label indicates the correct joint, then click to pick the joint. If not, click the right mouse button to shuffle the overlapping joints until a label indicating the correct joint appears, then click. After picking the pin joint the ring gear will be highlighted. Note that a rotation arrow (double arrowhead in pink color) appears that is consistent with the direction of the joint, as shown in Figure 7-15a. Also, make sure that *body3* (ring gear) is *Pinion*, and *Ground* is *Carrier*. Enter 8.167 for *Pitch Circle Diameter*.

Click the *Gear2* tab to define the second gear in the pair. Pick the pin joint *Pin2* (between the planet gear and the arm), and enter 2.333 for *Pitch Circle Diameter*. Note that the default direction is opposite to that of *Gear1*. We need to click the *Flip* button 🔲 (see Figure 7-15b) to flip the direction so that the planet gear and the ring gear rotate in the same direction, as it is supposed to be. Click *OK*. A gear symbol will appear between the two pin joints chosen (see Figure 7-16).

Now we will create a driver at joint *Pin1* to drive the arm, therefore the gear train system.

Click the *Servo Motors* button 🔲 of the *Insert* group on top of the *Graphics* window; a new set of selections will appear (like those shown in the previous lessons, for example, Figure 6-23). Pick *Pin1* from the *Graphics* window. After picking the pin joint, larger arrows appear to confirm your selection. The rotation axis must point in a direction like that of Figure 7-16. You may need to click the *Flip* button to make sure you pick the correct direction of the rotation axis. Click the *References* tab to make sure that *Pin1.first_rot_axis* has been selected for *Driven Entity*.

The next step is to specify the profile of the motor. Click the *Profile Details* tab (next to the *References* tab), choose *Angular Velocity* for *Driven Quantity* and *Constant* (default) for *Function Type*. Enter *360* for the constant *A* under *Coefficients*.

Click the *Properties* tab to enter *Motor1_arm* for name. Click the ✅ button on the right to accept the definition. A motor symbol appears in the *Graphics* window overlapping with *Pin1*, as shown in Figure 7-17.

Repeat the process to define the second motor. Click the *Servo Motors* button 🔲, then pick *Pin3*. Click the *Profile* tab, leave *Angular Position* and *deg* (default) in *Driven Quantity*, and leave *Constant* (default) for *Function Type* under *Motor Function*. Leave *0* for the constant *A*. Click the *Properties* tab to enter *Motor2_ring_gear* for name. Click the ✅ button on the right to accept the definition. Again, a driver symbol should appear at the pin joint *Pin3* (overlapping with *Pin1*) in the *Graphics* window.

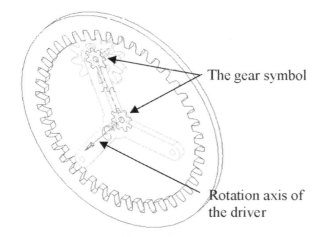

Figure 7-16 The Driver Direction Axis

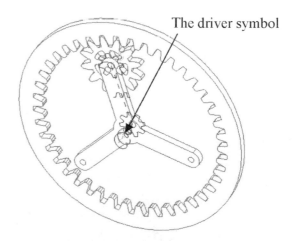

Figure 7-17 The Gear and Driver Symbols

Creating and Running a Kinematic Analysis

Click the *Mechanism Analysis* button 🔲 to define an analysis. In the *Analysis Definition* dialog box appearing, leave the default name, *AnalysisDefinition1*, and enter:

Type: *Kinematic*
Start Time: *0*
End Time: *1* (to drive the arm for a full cycle)
Frame Rate: *100*
Minimum Interval: *0.01*
Initial Configuration: *Current*

Make sure that both motors are included in the analysis. Run the analysis; in the *Graphics* window, the gears should start turning. The arm rotates 360 degrees counterclockwise as expected, driving the planet gear, and the teeth between the planet and ring gears mesh well.

Saving and Reviewing Results

Click the *Playback* button ◀▶ of the *Analysis* group to bring up the *Playbacks* dialog box and repeat the motion animation. From the *Playbacks* dialog box, click the *Save* button 💾 to save the results as a *.pbk* file. Play the animation. Are gear teeth properly meshed for the entire time period? They are. The motion of the gear train seems to be working correctly. The arm rotates in a counterclockwise direction, the planet gear rotates clockwise, and the ring gear is stationary.

Note that we want to create a measure to monitor the output angular velocity of the planet gear. We will choose joint *Pin2* and the angular velocity of its rotational axis for the measure. To do so, click the *Measures* button ⬜ of the *Analysis* group. In the *Measure Results* dialog box appearing, click the *Create New Measure* button 🗋. The *Measure Definition* dialog box opens. Enter *Planet_Gear_Velocity* for *Name*. Under *Type*, select *Velocity*. Pick *Pin2* in the *Graphics* window. Under *Evaluation Method*, leave *Each Time Step*. Click *OK* to accept the definition.

In the *Measure Results* dialog box, choose *AnalysisDefinition1* in the *Result Set* and click the *Graph* button 📈 at the top left corner to graph the measure. The graph should be similar to that of Figure 7-18, which shows that the angular velocity of the planet is a constant of 1,260 degrees/sec. Note that this value is different from what we calculated on page 7-3, i.e., –900 degrees/sec. Although the motion animation seems to be correct, unfortunately, in this case, *Mechanism* gave wrong result graphs. Basically, *Mechanism* considers this gear train as a simple spur gear train, and calculated the velocity using equations discussed in *Lesson 6*; i.e.,

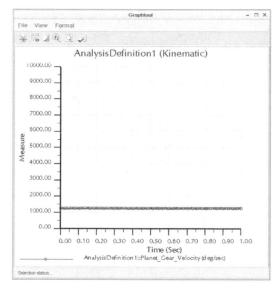

Figure 7-18 Angular Velocity of the Planet Gear

$$\omega_{out} = \frac{N_r}{N_p}\omega_{in} = \frac{42}{12}360 = 1,260 \qquad (7.4)$$

In fact, *Mechanism* does not support simulation for planetary gear train system. Users should first check the capabilities of the software before using it for simulating specific systems. Save your model before we move to the next example. Since the *Mechanism* does not support simulation for planetary gear train system, the next exercise should be treated as simply creating an animation for visual presentation.

Example 2: Multiple Planetary Gear Train

We will add a sun gear and two planet gears to the current model for the second example. The sun gear will be connected to the ground using a pin joint and will be driven by a motor. Note that in this example, the motor that drives the arm will be removed. The two additional planet gears will be carried by the arm, similar to that of *Example1*, in addition to being driven by the sun gear.

Click the *Applications* tab and click the *Mechanism* button 🔘 to exit from *Mechanism*. Click *Model* tab to go back to *Creo* assembly mode. Save the current model as *Example2* (use *File > Save As > Save a Copy*). In the *Save A Copy* dialog box, enter *Example2*, then click *OK*. In the *Assembly Save A Copy* dialog box that appears next, click *Save Copy*. Close the current model (using *File > Close*), and then remove not-displayed objects from current session by choosing *File > Manage Session > Erase Not Display*. Open *Example2*.

We will bring in the sun gear, *sun.prt*. Click the *Assemble* button 🔧 and choose *sun.prt*. In the *Component Placement* dashboard from the *User Defined* list choose the *Pin* joint, and pick *AA_1* (assembly) and *A_2* (*sun.prt*), as shown in Figure 7-19a. Now turn off datum axis display and turn on datum plane display. Pick two datum planes, *ASM_FRONT* (assembly) and *FRONT* (*sun.prt*), as shown in Figure 7-19b. The sun gear is not properly meshed with the planet gear yet (we will address this issue shortly). Similar to the steps in bringing the ring gear discussed earlier, click the *Placement* button, click the pull-down button and select *Distance* (see Figure 7-12c), and enter *–0.3125* for *Offset*. With this offset, the sun gear will be properly aligned with the planet gear. Change the joint name to *Pin4* and click ☑ to accept the definition.

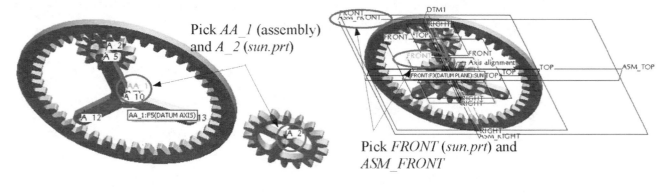

Pick *AA_1* (assembly) and *A_2* (*sun.prt*)

Pick *FRONT* (*sun.prt*) and *ASM_FRONT*

(a) Align Datum Axes   (b) Align Datum Planes

Figure 7-19 Assembling *sun.prt*

Now we repeat the same steps as we discussed earlier to assemble two additional planet gears to the arm. Basically, we define a pin joint between the planet gear and the arm by aligning axis *A_2* of the planet gear with *A_12* of the arm (and then repeat it for the second planet gear with *A_13* of the arm, as shown in Figure 7-20), and aligning the front face of the planet gear to the back face of the arm. Change the joint names to *Pin5* and *Pin6*, respectively. The complete assembly should be like the one shown in Figure 7-21.

Now we must check if all gears mesh properly. Hide the arm (right click *ARM.PRT* from the model tree and click the *Hide* button 👁). Use the *FRONT* view to see the four gears, such as in Figure 7-22. The gear teeth are not properly meshed. It is not too difficult to fix the problem. The only part that needs to be adjusted is the sun gear. The adjustment is to rotate the sun gear 90 degrees counterclockwise. This can be accomplished by aligning datum planes of the sun and the assembly using the *Drag* option. Before we rotate the sun gear, we will hide all other parts. Select all three planet gears and the ring gear from the model tree to hide them. You should see only the sun gear and six datum planes are displayed (three datum planes for sun gear and three for the assembly), as shown in Figure 7-23.

Now, click the *Drag Components* button 🖐. In the *Drag* dialog box, click the *Constraints* tab, and choose the *Align Two Entities* button (first on the left). Choose two datum planes *TOP* (*sun.prt*) and *ASM_RIGHT* (assembly), as shown in Figure 7-23. The sun gear is immediately rotated with these two datum planes aligned, and the gears should now mesh properly. Click the *Current Snapshot* button 📷 on top; a default name *Snapshot2* will appear. Click *Close*. The snapshot *Snapshot2* has been saved for future use. Now the gear train is properly assembled.

Unhide all gears (by right clicking the three planet gears and the ring gear in the model tree and choose *Unhide*). Choose the *FRONT* view; you should see that the gears are properly meshed, as shown in Figure 7-24. Now you can unhide the arm. Save your model before we enter *Mechanism*.

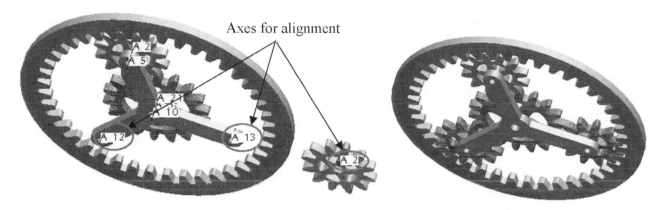

Figure 7-20 Axes for Alignment   Figure 7-21 The Complete Assembly

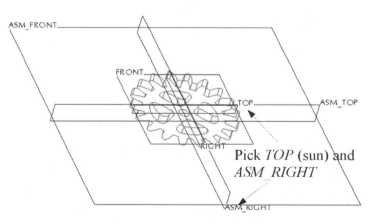

Figure 7-22 Gears Meshed Incorrectly   Figure 7-23 Sun Gear and Six Datum Planes

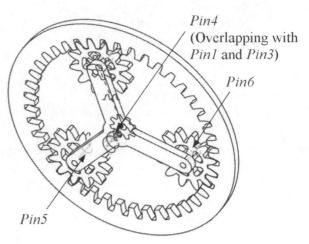

Pin4
(Overlapping with
Pin1 and Pin3)

Pin6

Pin5

Figure 7-24 The Gears Properly Meshed   Figure 7-25 Three Pin Joints Added

Enter *Mechanism*. There are additional three pin joints appearing, as shown in Figure 7-25. Note that *Pin4* overlaps with the other two pin joints *Pin1* and *Pin3* at the center of the gear train system. We will define a driver at *Pin4* (the joint between the sun gear and the ground).

First, we will define three more gear pairs: between sun gear and the three planet gears, respectively. Follow the same steps as discussed earlier. One thing to pay attention to is the fact that you must pick the right joint for the sun gear since there are now three joints that overlap at the center. The joint to pick is *Pin4*. The pitch circle diameter of the sun gear is 3.5. Recall that the pitch circle diameter of the planet gear is 2.333. Also note that you should not need to flip the direction arrow since the sun and planet gears rotate in the opposite direction.

Next, we will add two gear pairs between the two newly added planet gears and the ring gear. Recall that the pitch circle diameter of the ring gear is 8.167. The pin joint between the ring gear and the ground is *Pin3*. You may need to flip the direction arrow to ensure that the planet and ring gears rotate in the same direction.

Follow the same process as before to define the third motor. Pick *Pin4*. Click the *Profile* tab to define constant angular velocity of 360 degrees/sec. Click the *Properties* tab to enter *Motor3_sun_gear* for *Name*. Again, a driver symbol should appear at the pin joint *Pin4* (overlapping with *Pin1* and *Pin3*) in the *Graphics* window.

Now that the motion model is completed, we are ready to create and run a kinematic analysis. Note that we will delete the first motor (*Motor1_arm*) and include both *Motor2_ring_gear* (to keep the ring gear stationary) and *Motor3_sun_gear* (to drive the sun gear).

Expand the *ANALYSES* entity in the lower half of the model tree (which lists entities related to *Mechanism*). Right click *AnalysisDefinition1*, and choose *Edit Definition*. The *Analysis Definition* dialog box will be brought back. We will keep the same *Preferences* for the time being. Click the *Motors* tab; two motors should be listed (*Motor1_arm* and *Motor2_ring_gear*). Click the *Add All Motors* button (3rd on the right) to include all three motors. Click the first motor (*Motor1_arm*) and click the *Delete Highlighted Row(s)* button to delete the motor (2nd on the right). The second and third motors should be listed as shown in Figure 7-26.

Click the *Run* button to run the analysis. In the *Graphics* window, the mechanism should start moving. The sun gear rotates counterclockwise, which drives the planet gears rotating clockwise. The arm rotates counterclockwise due to the motion of planet gears at a lower speed than the sun gear. The planet gears rotate in the clockwise direction. All gears mesh well, and the animation correctly shows the kinematics of the planetary gear train.

Next, we define measures for the angular velocities of the sun gear (*Pin4*), the arm (*Pin1*), and the planet gear (any one of the three). Graph all three measures. You should see these graphs as shown in Figures 7-27 to 7-29. As shown in Figure 7-27, the angular velocity of the sun gear is 360 degrees/sec as expected (driven by *Motor3_sun_gear*). The angular velocities of the planet gear and arm are 378.1 degrees/sec and 108 degrees/sec, respectively, as shown in Figures 7-28 and 7-29.

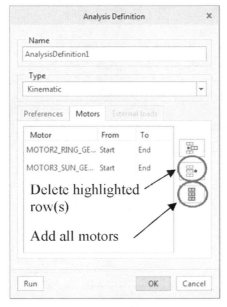

Figure 7-26 The *Analysis Definition* Dialog Box

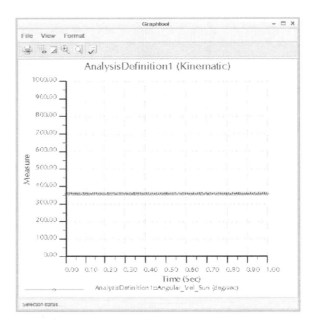

Figure 7-27 Angular Velocity of the Sun Gear

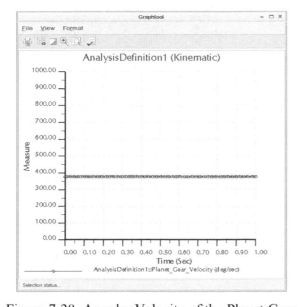

Figure 7-28 Angular Velocity of the Planet Gear

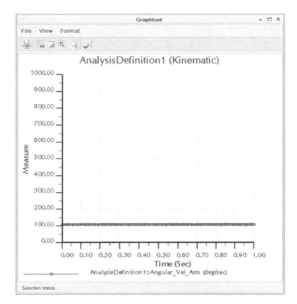

Figure 7-29 Angular Velocity of the Arm

Note that our calculations indicate that the arm will rotate only 108 degrees in the same direction as that of the sun gear, i.e., counterclockwise. However, the planet gears should turn 270 degrees per revolution of the sun gear in the clockwise direction, which is not the result indicated in Figure 7-28. Although gear motion shown in the animation looks good, *Mechanism* does not produce accurate measures. As demonstrated in this lesson, it is extremely important that you verify the simulation results before accepting them for any purposes.

Exercises:

1. Modify *Example1* (single planetary gear train) by removing the arm, allowing the ring gear to rotate, and keeping the planet gear stationary. Create a driver to rotate the planet gear at 360 degrees/sec counterclockwise; what is the angular velocity of the ring gear? Does *Mechanism* provide the correct answer in terms of animation and measure graph?

2. From *Example2* (multiple planetary gear train), change the rotation speed of the sun gear from 360 to −360 degrees/sec. Edit the kinematic analysis by removing *Motor2_ring_gear* and including *Motor1_arm*. The arm will be driven by the motor at 360 degrees/sec. Run the kinematic analysis. What is the angular velocity of the ring gear obtained from *Mechanism*? Is the result correct? Verify it using analytical calculation discussed in Section 7.2.

Notes:

Lesson 8: Cam and Follower

8.1 Overview of the Lesson

In this lesson, we will learn cam and follower, or cam-follower connection. A cam-follower is a connection that converts rotary motion into linear motion of desired characteristics. The simplest form of a cam is a rotating disc with a variable radius, so that its profile is not circular but oval or egg-shaped. When the disc rotates, its edge pushes against a follower (or cam follower), which may be a small wheel at the end of a lever or the end of the rod itself. The follower will thus rise and fall at exactly the same amount as the variation in cam radius. By profiling a cam appropriately, a desired cyclic pattern of linear or straight-line motion, in terms of position, velocity, and acceleration, can be produced.

We will learn to create a motion model to simulate the control of opening and closing of an inlet or exhaustive valve, usually found in internal combustion engines, using cam-follower connections. In a design such as that of Figure 8-1, the drive for the camshaft is taken from the crankshaft through a timing chain, which keeps the cams synchronized with the movement of the piston so that the valves are opened or closed at precise instants. The mechanism we will be working with consists of a camshaft, pushrod, rocker, valve, and spring, as shown in Figure 8-1. There are three pairs of cam-followers connecting the camshaft and the pushrod, the pushrod and the rocker, and the rocker and the valve. When the cam on the camshaft pushes the pushrod up, the rocker rotates and pushes the valve on the other side downward. The spring surrounding the valve gets compressed and opens up the inlet for air to flow into (or out of) the combustion chamber.

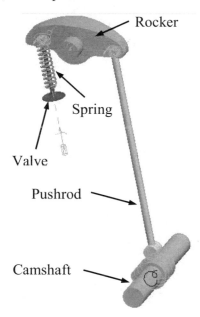

Figure 8-1 The Mechanism of Engine Inlet or Outlet Valve

8.2 The Cam and Follower Example

Physical Model

The camshaft and the rocker will rotate along their respective pin joints connecting them to the ground body. The camshaft is driven by a motor of constant velocity of 600 rpm. The profile of the cam consists of two circular arcs of 0.25 and 0.5 in. radii, respectively, as shown in Figure 8-2. The lower arc is concentric with the shaft, and the center of the upper arc is 0.52 in. above the center of the shaft. When the camshaft rotates, the cam mounted on the shaft pushes the pushrod up by up to 0.27 in. (that is, 0.52+0.25−0.5 = 0.27). As a result, the rocker will rotate and push the valve at the other end downward by 0.27 in. at a frequency of 10 Hz. When the camshaft rotates where the larger circular arc (0.5 in. radius) of the cam is in contact with the follower (the pushrod), the pushrod has room to move downward. The rocker will rotate clockwise since the spring is being uncompressed. As a result, the valve will move up, and therefore, close the inlet. The valve will be open for about 120 degrees per cycle.

The unit system chosen for this example is in-lb$_f$-sec and all parts are made of steel.

Creo Parts and Assembly

The cam and follower example consists of one assembly and four parts: *partial_cam_follower.asm*, *cam_shaft.prt*, *rocker.prt*, *pushrod.prt*, and *valve.prt*. You may want to open the final assembly, *cam_follower.asm*, to check the assembled cam and follower example shown in Figure 8-1 before going over this lesson. Enter *Mechanism* (by choosing *Applications > Mechanism*). Choose *Analysis > Playback* to show the motion. This is what we want to accomplish in this lesson.

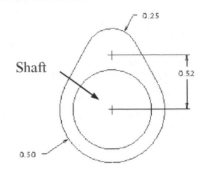

Figure 8-2 The Cam Profile

The *partial_cam_follower.asm* is where we will start in this lesson. This partial assembly consists of three datum planes and a datum coordinate system by default. In addition, there are four datum axes (*AA_1* to *AA_4*) and one datum point *APNT0* for assembling components, as shown in Figure 8-3.

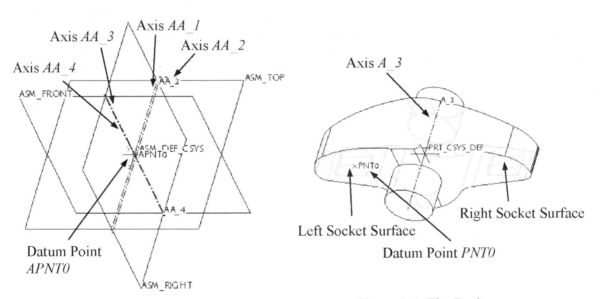

Figure 8-3 The Partial Assembly

Figure 8-4 The Rocker

The datum axis *A_3* in the rocker shown in Figure 8-4 will be used to create a pin joint between the rocker and the ground body (axis *AA_1* in assembly). The datum point *PNT0* is on the bottom face of the rocker close to the left socket, which will be used for creating a spring (connecting datum point *APNT0*) that wraps around the valve once it is assembled. The two socket surfaces, which are semi-circular cylindrical surfaces, will be used to assemble the pushrod (with *Right Socket Surface*) and the valve (with *Left Socket Surface*), respectively, using cam-follower connections.

The datum axis *A_14* in the pushrod (Figure 8-5) will be used to align with assembly datum axis *AA_4*. In addition, the datum plane *FRONT* of the pushrod will be aligned with assembly datum plane *ASM_FRONT* for defining its orientation. The circular cylindrical surface *Cylindrical Surface 1* (shown in Figure 8-5) will be assembled to the camshaft (with *Cam Surface* of the camshaft shown in Figure 8-6) using a cam-follower connection. The second cylindrical surface *Cylindrical Surface 2* at the top of the pushrod will be assembled to the rocker (with *Right Socket Surface* of the rocker shown in Figure 8-4) using a cam-follower connection.

The datum axis *A_2* in the camshaft shown in Figure 8-6 will be used to create a pin joint between the camshaft and the ground body (align with datum axis *AA_2*). The cam surface, with the profile shown in Figure 8-2, will be used to create a cam-follower with *Cylindrical Surface 1* of the pushrod (Figure 8-5).

The datum axis *A_6* in the valve (Figure 8-7) will be used to create a cylinder joint with assembly datum axis *AA_3*. The circular cylindrical surface *Cylindrical Surface 3* at the top will be assembled to the rocker (with *Left Socket Surface* of the rocker shown in Figure 8-4) using a cam-follower connection. The datum point *PNT0* is for creating measures that monitor the valve motion.

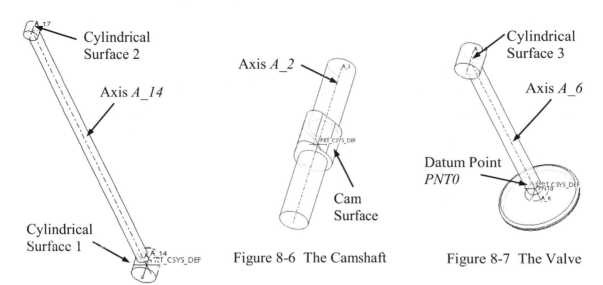

Figure 8-5 The Pushrod

Figure 8-6 The Camshaft

Figure 8-7 The Valve

Simulation Model

The simulation model shown in Figure 8-8 consists of two pin joints, one cylinder joint, three cam-followers, a spring, and a servo motor. The servo motor rotates 600 rpm (3,600 degrees/sec) that drives the mechanism. The spring surrounding the valve has a spring constant of 10 lb$_f$/in and a free length of 1.45 in., which is the vertical distance (along the *Y*-direction of *ASM_DEF_CSYS*) between the two datum points, *APNT0* of the assembly and *PNT0* of the rocker. Note that cam-followers 1 and 3 simply connect pushrod to rocker and rocker to the valve, respectively. The connecting surfaces are circular cylindrical surfaces of constant radii. Therefore, they are not the typical cam-follower like that of cam-follower 2.

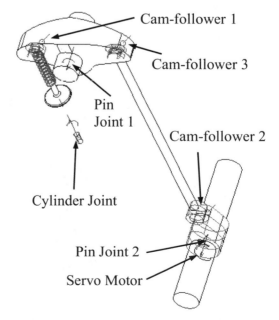

Figure 8-8 The Simulation Model

8.3   Using *Mechanism*

Creating an Assembly

Start *Creo*, select working directory, and open *partial_cam_follower.asm*. You should see assembly datum features in the *Graphics* window. You may want to save the assembly to a different name.

First, we will bring in *rocker.prt*. Click the *Assemble* button and choose *rocker.prt*. In the *Component Placement* dashboard from the *User Defined* list, choose *Pin* joint. Turn on datum axis display and pick *AA_1* (*partial_cam_follower.asm*) and *A_3* (*rocker.prt*), as shown in Figure 8-9a. Now turn off datum axis display and turn on datum plane display. Pick two datum planes, *ASM_FRONT* (*partial_cam_follower.asm*) and *FRONT* (*rocker.prt*), as shown in Figure 8-9b. The *rocker.prt* should be properly assembled through a pin joint. Click to accept the definition.

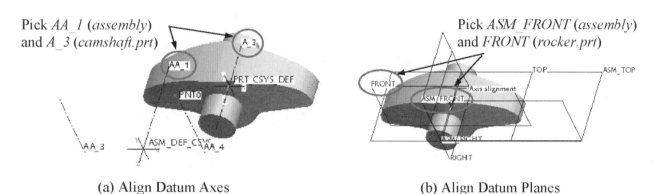

Pick *AA_1* (*assembly*) and *A_3* (*camshaft.prt*)

Pick *ASM_FRONT* (*assembly*) and *FRONT* (*rocker.prt*)

(a) Align Datum Axes   (b) Align Datum Planes

Figure 8-9 Assembling *rocker.prt*

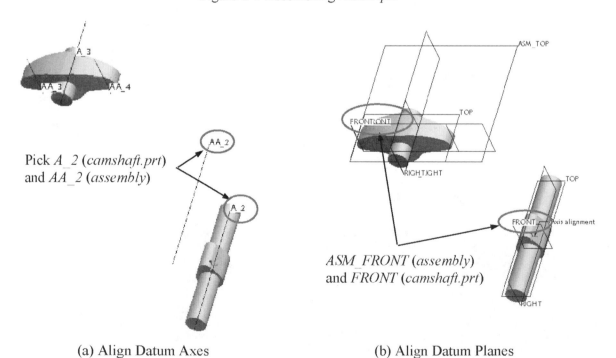

Pick *A_2* (*camshaft.prt*) and *AA_2* (*assembly*)

ASM_FRONT (*assembly*) and *FRONT* (*camshaft.prt*)

(a) Align Datum Axes   (b) Align Datum Planes

Figure 8-10 Assembling *camshaft.prt*

The next part we will bring in is the camshaft. Click the *Assemble* button and choose *camshaft.prt*. Choose the *Pin* joint, turn on datum axis display, and pick *A_2* (*camshaft.prt*) and *AA_2* (*partial_cam_follower.asm*), as shown in Figure 8-10a. Now turn off datum axis display and turn on datum plane display. Pick two datum planes, *ASM_FRONT* (*partial_cam_follower.asm*) and *FRONT* (*camshaft.prt*), as shown in Figure 8-10b. The *camshaft.prt* should be properly assembled to the ground body with a pin joint. Click ✓ to accept the definition.

The third part we bring in is *valve.prt*. Choose the *Cylinder* joint (from *User Defined* list), turn on datum axis display, and pick *AA_3* (*partial_cam_follower.asm*) and *A_6* (*valve.prt*), as shown in Figure 8-11. The valve is free to move along the axis. We will leave it as it is for the time being. We will later define a cam-follower in *Mechanism* to properly position the valve to the rocker. For the time being, click ✓ to accept the definition.

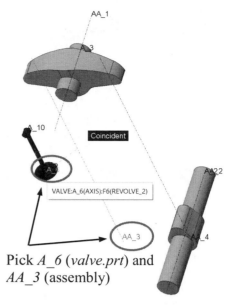

Pick *A_6* (*valve.prt*) and *AA_3* (assembly)

Figure 8-11 Assembling *valve.prt*

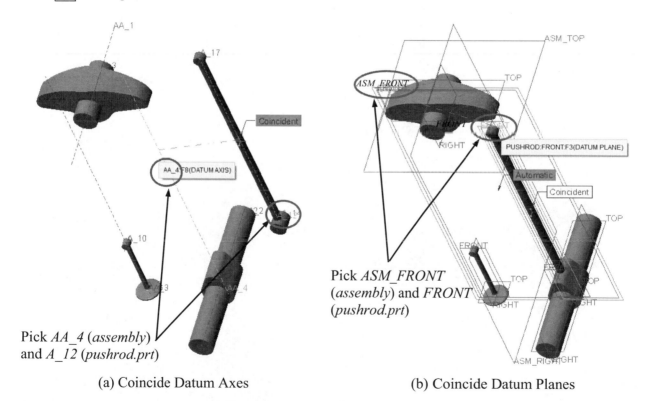

Pick *AA_4* (*assembly*) and *A_12* (*pushrod.prt*)

Pick *ASM_FRONT* (*assembly*) and *FRONT* (*pushrod.prt*)

(a) Coincide Datum Axes   (b) Coincide Datum Planes

Figure 8-12 Assembling *pushrod.prt*

The fourth and final part we will bring in is *pushrod.prt*. Choose *Coincident* (from *Automatic* list), and pick *AA_4* (*partial_cam_follower.asm*) and *A_12* (*pushrod.prt*), as shown in Figure 8-12a. Choose *Coincident* again (note that you will have to choose *Placement* tab, click *New Constraint*, and choose *Coincident*), and pick *ASM_FRONT* (*partial_cam_follower.asm*) and *FRONT* (*pushrod.prt*), as shown in Figure 8-12b. Similar to the valve, we will later define a cam-follower in *Mechanism* to properly position the pushrod between the rocker and the camshaft. Now the pushrod is partially constrained. It is fine since we will add cam-follower connections later. Click ☑ to accept the definition.

The model is now mostly assembled. Save your model and move to *Mechanism*.

Creating a Simulation Model

Choose *Applications > Mechanism* to enter *Mechanism*.

There are three connection symbols appearing: two pin joints and one cylinder joint, as shown in Figure 8-13. We will choose three pairs of surfaces to define the three cam-followers.

From the *Connection* group on top of the *Graphics* window, click *Cam* button 🔘. The *Cam-Follower Connection Definition* dialog box appears (Figure 8-14). Use the default name (*CamFollower1*), click *Autoselect* (to select all surfaces surrounding the cam or follower), and click the *Select* button 🔲. Pick the cylindrical surface (*Cylindrical Surface 3*) on top of the valve; all the surfaces surrounding the cylinder will be selected, as shown in Figure 8-15a. Click the *OK* button in the *Select* dialog box (right underneath the *Cam-Follower Connection Definition* dialog box). The surface selected will appear in the *Surfaces/Curves* text area (see Figure 8-14). An arrow will also appear on the surfaces selected showing their normal. The normal vector should be pointing outward.

Click the *Cam2* tab and repeat the same process but select the inner cylindrical surface of the rocker (*Left Socket Surface*), as shown in Figure 8-15a. Make sure the normal vector is pointing outward. You should not need to flip the normal vector; but if needed, you may click the *Flip* button (Figure 8-14) to flip it. Click *OK* in the *Cam-Follower Connection Definition* dialog box. A cam-follower symbol ⟲ will appear between the two surfaces selected.

Repeat the same process and pick the two surfaces shown in Figure 8-15b for the second cam-follower between the pushrod and the camshaft. Finally, pick the two surfaces to create the third cam-follower connection between the rocker and the pushrod (Figure 8-15c).

Now we will create a servo motor at pin joint 2 (between the camshaft and the ground body). The servo motor will drive the camshaft at an angular velocity of 3,600 degrees/sec, therefore, the whole system.

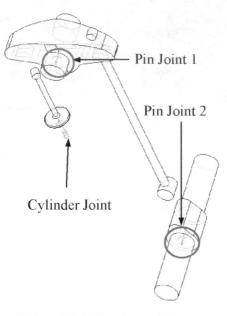

Figure 8-13 The Assembly Entering *Mechanism*

Figure 8-14

Click *Servo Motors* button 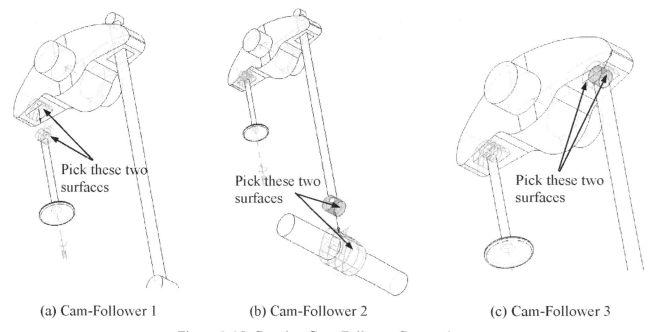 of the *Insert* group on top of the *Graphics* window; a new set of selections will appear (see Figure 8-16). Pick the pin joint of the camshaft from the *Graphics* window. Note that you may want to turn off all datum features in order to see the pin joint. After picking the pin joint, a larger arrow appears to confirm your selection. The rotation axis must point in a direction like that of Figure 8-16. Click the *References* tab to make sure that *Connection_48.first_rot_axis* has been selected for *Driven Entity*, as circled in Figure 8-16. Next, click the *Profile Details* tab, choose *Angular Velocity* for *Driven Quantity*, and leave *Constant* (default) for *Function Type* under *Motion Function*. Enter *3600* for the constant *A* (3,600 degrees/sec, i.e., 600 rpm). Click the *Properties* tab to enter *Motor1* for name.

Pick these two surfaces

Pick these two surfaces

Pick these two surfaces

(a) Cam-Follower 1   (b) Cam-Follower 2   (c) Cam-Follower 3

Figure 8-15 Creating Cam-Follower Connections

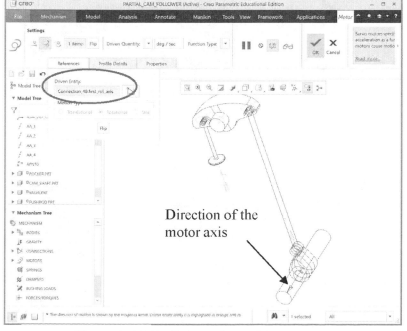

Direction of the motor axis

Figure 8-16 Defining a Servo Motor

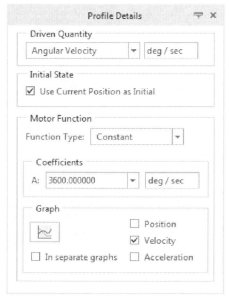

Profile Details

Driven Quantity

Angular Velocity   deg / sec

Initial State

☑ Use Current Position as Initial

Motor Function

Function Type: Constant

Coefficients

A: 3600.000000   deg / sec

Graph

☐ Position
☑ Velocity
☐ In separate graphs   ☐ Acceleration

Figure 8-17

Click the ✓ button on the right to accept the definition. A motor symbol appears in the *Graphics* window overlapping with the pin joint, as circled in Figure 8-18.

Next, we create a spring surrounding the valve. The spring will be created by connecting two datum points: *PNT0* (*rocker*) and *APNT0* (assembly), as shown in Figure 8-18. Note that *APNT0* is located at (−1.25, −1.75, 0) from the assembly coordinate system *ASM_DEF_CSYS*. *PNT0* is located at the bottom face of the rocker. It is 1.25 in. from the *RIGHT* datum plane and is 0.3 in. below axis *A_3* (that passes through the origin of *ASM_DEF_CSYS*). Therefore, these two points align vertically and are 1.45 in. apart (along the *Y*-direction of *ASM_DEF_CSYS*). The free length of the spring should be about 1.45 in., assuming that the rocker does not rotate.

Click the *Springs* button 🌀 of the *Insert* group; a new set of selections will appear at the top of the *Graphics* window for defining the spring (Figure 8-19). Choose *Extension or compression spring* button ⊣ (the first button from the left; it should have been selected by default). Activate the *Select items* field by clicking it. Turn on the datum point display. Then, pick *PNT0* of the rocker. Drag the handle appeared in the *Graphics* window (Figure 8-20) and overlap it with *APNT0* (release the mouse button when *APNT0* is highlighted). A spring will appear, connecting *PNT0* (*rocker*) and *APNT0* (ground).

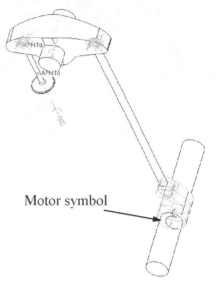

Motor symbol

Figure 8-18 Motor Added to Pin Joint 2

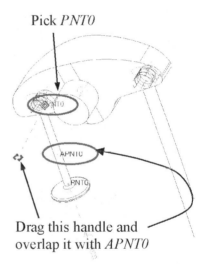

Pick *PNT0*

Drag this handle and overlap it with *APNT0*

Figure 8-20 Defining Spring

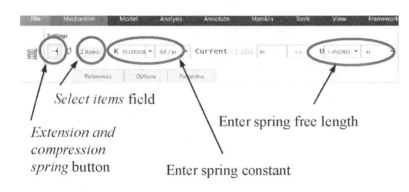

Select items field

Extension and compression spring button

Enter spring free length

Enter spring constant

Figure 8-19 The *Spring Definition* Field

Next, enter 10 (lb$_f$/in) for spring constant (**K**) and 1.45 (in.) for the free spring length (**U**) from the text fields at the top of the *Graphics* window. Note that the spring constant we enter is in lb$_f$/in units. Click ✓ at right to accept the definition. A spring symbol should appear in the *Graphics* window.

Now the mechanism is completely defined. You may click the *Drag Components* button 🖐 at the top of the *Graphics* window, and click the camshaft. You should be able to move (rotate) the camshaft along the pin joint, and therefore, drive the whole mechanism. The rocker will rotate and push the valve and compress the spring. You may want to save the model before defining a motion analysis.

Creating and Running a Static Analysis

We would like to start with an equilibrium configuration for dynamic analysis. An equilibrium configuration can be obtained by conducting a static analysis. Click the *Mechanism Analysis* button ⊠ to define an analysis. In the *Analysis Definition* dialog box appearing (Figure 8-21), enter *Static_Analysis* for *Name*, and choose *Static* for *Type*. Click *Run*. The static analysis will start and a *Graphtool* window similar to Figure 8-22 will appear showing the progress of the analysis. Note that the graph you have may be slightly different from that of Figure 8-22, depending on the configuration of the mechanism you currently have. In the *Graphics* window, the mechanism will situate to an equilibrium configuration where the rocker stays almost leveled and the spring is in its (almost) free length, as shown in Figure 8-23. Choose the *FRONT* view to see the equilibrium configuration.

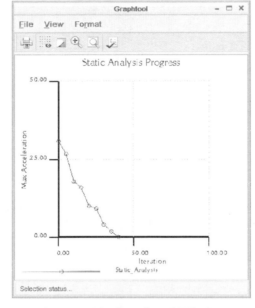

Figure 8-22. *Graphtool* Window

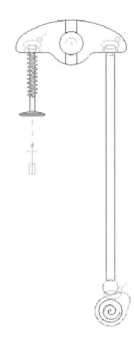

Figure 8-21

Figure 8-23 An Equilibrium Configuration

Use the *Drag Components* button 🖑 to create a snapshot of the current configuration. We will use the snapshot as an initial condition for the dynamic analysis.

Creating and Running a Dynamic Analysis

Click the *Mechanism Analysis* shortcut button ⊠ to define a dynamic analysis. In the *Analysis Definition* dialog box appearing, enter:

Name: *Dynamic_Analysis*
Type: *Dynamic*
Duration: *0.5* (for the camshaft to rotate 5 full cycles)
Frame Rate: *1000*
Minimum Interval: *0.001*
Initial Configuration: *Current*

Make sure that *Motor1* is included. Run the analysis. In the *Graphics* window, the mechanism should start moving. The camshaft rotates 5 complete cycles counterclockwise.

Saving and Reviewing Results

Click the *Playback* button ◀▶ to bring up the *Playbacks* dialog box and play the motion animation. On the *Playbacks* dialog box, choose *Dynamic_Analysis* for *Result Set*, and then click on the *Save* button 💾 to save the dynamic analysis results as a *.pbk* file.

Note that we want to create a measure to monitor the position of the valve. We will choose the *Y*-position of the datum point *PNT0* of the valve (see Figure 8-24) for the measure. The graph of the position measure is shown in Figure 8-25, where the valve is moving between −2.45 and −2.15, traveling about 0.3 in.

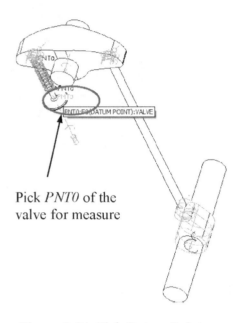

Pick *PNT0* of the valve for measure

Figure 8-24 Pick Datum Point

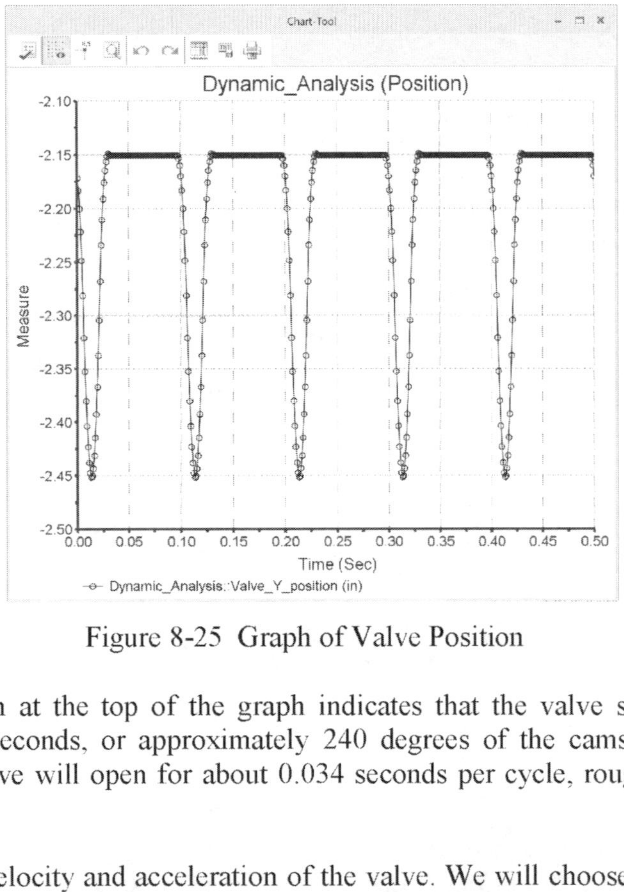

Figure 8-25 Graph of Valve Position

As shown in Figure 8-25, the flat portion at the top of the graph indicates that the valve stays completely closed, which spans about 0.066 seconds, or approximately 240 degrees of the camshaft rotation in a complete cycle. Therefore, the valve will open for about 0.034 seconds per cycle, roughly 120 degrees.

Next, we create measures to monitor the velocity and acceleration of the valve. We will choose the *Y*-velocity and acceleration of the datum point *PNT0* of the valve for the measures. The graphs of the velocity and acceleration measures are shown in Figures 8-26 and 27, respectively. As shown in Figure 8-26, there are two velocity spikes per cycle, representing the valve being pushed downward (negative velocity) for opening and being pulled back (positive velocity) for closing, respectively. The valve stays closed with zero velocity.

Figure 8-27 reveals high accelerations when the valve is pushed and pulled. Note that such high acceleration is due to high-speed rotation at the camshaft. However, these high accelerations could produce large inertial force on the valve, yielding high contact force between the top of the valve and the socket in the rocker. The reaction force can be monitored by defining a reaction measure for *Cam-*

Follower 1 (see Figure 8-8 for its location). Choose *Normal Force* under *Component*, as shown in Figure 8-28. The reaction force graph (Figure 8-29) shows that the reaction force between the top of the valve and the socket face of the rocker is about 700 lb$_f$, which is significant. Note that if the camshaft rotates at a higher speed, e.g., 6,000 rpm, the reaction force would be around 7,000 lb$_f$, which raises a flag on the durability of the valve. In order to check the structural integrity of the valve under such a large force, a finite element analysis is usually conducted, e.g., using *Creo Simulate*.

The large reaction force looks suspicious. How does this small engine valve produce such a large force? If the force produced by the valve is indeed this large, the rocker would have been damaged quickly. Before moving to *Creo Simulate*, it is worthwhile to investigate further on the reaction force.

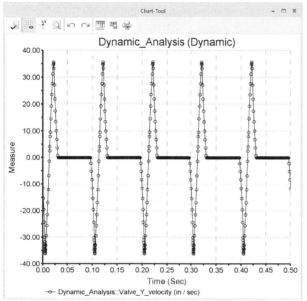

Figure 8-26 Graph of Valve Velocity

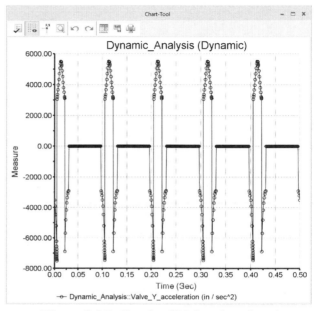

Figure 8-27 Graph of Valve Acceleration

Figure 8-28

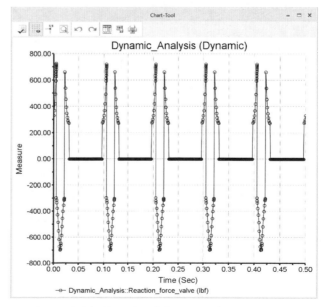

Figure 8-29 Graph of the Reaction Force

It turns out that the problem is that the mass property was not specified properly for the valve. If you open *valve.prt*, choose the *Analysis* tab and click *Mass Properties*; you should see a *Mass Properties* dialog box appears, similar to Figure 8-30. Pick the default coordinate system (*PRT_CSYS_DEF*) from the *Graphics* window; mass properties of the valve will appear in the box. It shows that the density is 1.0 lb_f s^2/in^4 (note that the mass unit in in-lb_f-sec unit system is lb_f s^2/in^4); hence, a total mass is close to 0.1 lb_f s^2/in (more accurately, 0.09397). With acceleration close to 6,000 in/s^2, the inertial force is about 600 lb_f. Note that the mass density is un-realistically large.

If steel is assigned to the valve as material, the total mass becomes 6.883×10^{-5} lb_f s^2/in, the inertia (therefore the reaction force) will be in the neighborhood of 0.4 lb_f, which is more manageable and much more realistic. It is critical that you set up a correct motion model in order to expect accurate simulation results. As shown in this lesson, it is more important that whenever results are suspicious, you will have to check the models very carefully, identify possible modeling errors, correct them, and carry out simulation again for results that you can trust.

Figure 8-30

Exercises:

1.   Redesign the cam by reducing the small arc radius from 0.25 to 0.2 and reducing the center distance of the small arc from 0.52 to 0.40, as shown in Figure E8-1. Repeat the dynamic analysis and check reaction force between the valve and the rocker. Does this redesigned cam alter the reaction force?

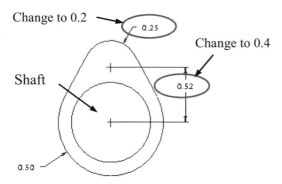

Figure E8-1 The Cam Profile

Notes:

Lesson 9: Assistive Device for Wheelchair Soccer Game

9.1 Overview of the Lesson

This is the first of the two application lessons. We will apply what we learned so far to real-life applications. The application of this lesson involves designing a customized mechanism that can be mounted on a wheelchair to mimic soccer ball-kicking action while being operated by a child with limited mobility and arm strength sitting on a wheelchair. Such a mechanism will provide more incentive and a realistic experience for children with physical disabilities to participate in a soccer game. This example was extracted from an undergraduate student design project that was carried out in conjunction with a local children's hospital. This device was intended primarily to be used in the summer camp sponsored by the children's hospital.

9.2 The Assistive Device

Physical Model

This assistive device for soccer playing consists of five major components: the clamper, handle bar, plate, kicking rod, and spring, as illustrated in Figure 9-1. In reality these five components will be assembled first and clamped to the lower frame of the wheelchair for use.

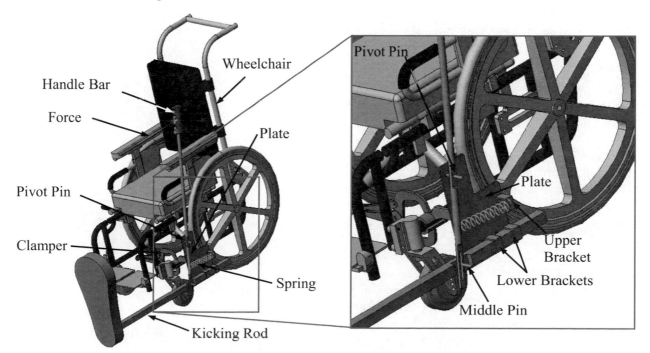

Figure 9-1 Assistive Device for Soccer Game

The handle bar is pinned at the pivot pin of the plate and linked to the middle pin of the kicking rod. The kicking rod is inserted into the two lower brackets mounted on the plate. When the handle bar is pulled backward, the handle bar rotates along the pivot pin, therefore driving the kicking rod to move forward along the longitudinal direction through the link between the end slot of the handle bar and the middle pin of the kicking rod. The forward movement produces momentum to "kick" the soccer ball. A spring is added between the upper bracket and the handle bar to restore the handle bar to its upright position after pulling.

The goal of this lesson is to use *Mechanism* to calculate the position, velocity, and acceleration at the kicking rod for a given force. With the simulation results, the design can be adjusted to ensure that a minimum pulling force will produce a sufficient momentum that drives the kicking rod to realistically mimic a soccer ball kicking. The design adjustment could include the location of the pivot pin on the handle bar, the location of the spring, the spring constant, etc.

Creo Parts and Assembly

All *Creo* parts are provided for this lesson. In addition, all parts are assembled, except for the handle bar and the kicking rod. These two components will be assembled with proper joints that allow the mechanism to move as desired. Note that the spring will be added to the system in *Mechanism*.

Table 9-1 List of Parts and Assemblies in Lesson 9 Folder

Assembly	Parts	Remarks
wheelchair_partial.asm		Partial assembly to start this lesson
	wheelchair.prt	Wheelchair part, including the clamper
	plate.prt	Plate with brackets
handle.prt		To be assembled
kicking_rod.asm	*rod.prt*	To be assembled
	foot.prt	

The example files you downloaded from the publisher's website should consist of five parts and two assemblies, as listed in Table 9-1. In addition, a completely defined motion model with analysis files (*Static.pbk* and *Dynamic.pbk*) is included for your reference. The in-lb$_f$-sec unit system has been chosen for all parts and assemblies.

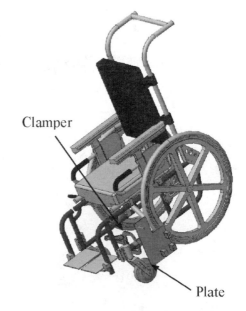

Note that the *kicking_rod.asm* and *handle.prt* are the only two components to be assembled to the *wheelchair_partial.asm*, which is the assembly we will use to start this lesson. Note that *wheelchair_partial.asm* consists of a wheelchair part and the clamper (filename: *wheelchair.prt*) with a plate (filename: *plate.prt*) attached, as shown in Figure 9-2.

In assembling these two components, we will use component placement constraints. The *kicking_rod.asm* will be first assembled to *plate.prt* using coincident placement constraint (aligning axis *A_6* of *plate.prt* and axis *A_3* of *rod.prt*). The axis coincident constraint will place the rod through the two lower brackets in the plate, as shown in Figure 9-3a.

Figure 9-2 The Partial Assembly

The rod will also be oriented with the foot at its end pointing upward, using a plane coincident constraint (see Figure 9-3b). Therefore, the kicking rod will be left with only one degree of freedom enabling it to slide along the longitudinal direction.

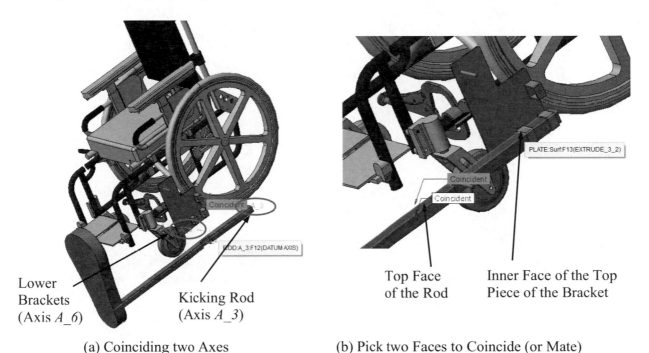

Lower Brackets (Axis *A_6*)   Kicking Rod (Axis *A_3*)

Top Face of the Rod   Inner Face of the Top Piece of the Bracket

(a) Coinciding two Axes   (b) Pick two Faces to Coincide (or Mate)

Figure 9-3 Assembling the Kicking Rod

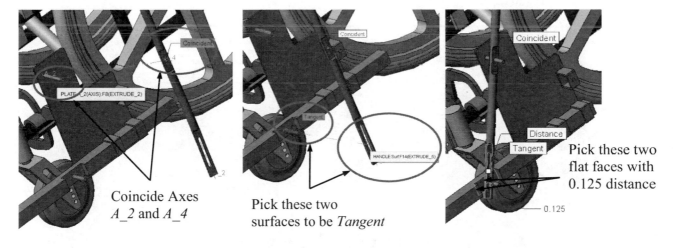

Coincide Axes *A_2* and *A_4*

Pick these two surfaces to be *Tangent*

Pick these two flat faces with 0.125 distance

(a) Coincide Two Axes   (b) Pick Two Tangent Surfaces   (c) Coincide Flat Faces

Figure 9-4 Assembling the Handle Bar

The handle bar is assembled to the kicking rod using an axis coincident, surface tangent, and surface coincident placement constraints. The axis coincident constraint will be defined between the handle bar and the plate by aligning two axes: *A_2* of *plate.prt* and *A_4* of *handle.prt*, as shown in Figure 9-4a. The tangent placement constraint will be defined by choosing the outer cylindrical surface of the middle pin of the kicking rod and the inner surface of the end slot of the handle bar, as shown in Figure 9-4b. In addition, the recessed flat face of the kicking rod where the middle pin resides and the back flat surface of the handle bar (close to its end where the tangent constraint was defined) coincide with a distance 0.125

in. (using the *Distance* placement constraint), as shown in Figure 9-4c. As a result, the handle bar is able to rotate along the pivot pin and drive the kicking rod to move along the longitudinal direction through the surface tangent constraint between the end slot and the middle pin.

Simulation Model

In this example, the partial assembly *wheelchair_partial.asm* is assigned as the ground body. The pin joint at the pivot pin will allow a rotational motion between the handle bar and the ground body, as shown in Figure 9-1. The kicking rod assembly will slide along a longitudinal direction. A spring will be added to restore the vertical orientation of the handle bar after pulling. Finally, an impulse force of 5 lbs in a time span of 0.5 seconds will be added to the handle bar to simulate the pulling force.

9.3 Using *Mechanism*

Assembling the Handle Bar and Kicking Rod

Start *Creo*, select working directory, and open the assembly: *wheelchair_partial.asm*. You should see *wheelchair_partial.asm* appear in the *Graphics* window like that of Figure 9-2. Note that the partial assembly consists of two components: the wheelchair (with clamper), and the plate (with brackets). They are listed in the model tree window, i.e., *WHEELCHAIR.PRT* and *PLATE.PRT*. Make sure the unit system has been set to in-lb$_f$-sec.

Refer to the discussion of Figures 9-3 and 9-4 to assemble the kicking rod and handle bar. Thereafter, click the *Drag Components* button at the top of the *Graphics* window, and click the handle bar. You should be able to move (rotate) the handle bar along the pivot pin and therefore drive the kicking rod back and forth along the longitudinal direction, as shown in Figure 9-5.

We will use the *Drag Components* button to create a snapshot for initial condition. The snapshot can be created by orienting (choosing the third button: *Orient two surfaces*) two surfaces, e.g., *RIGHT:F1(DATUM PLANE): HANDLE* and *ASM_FRONT* of the wheel chair assembly, as shown in Figure 9-7. Save the snapshot.

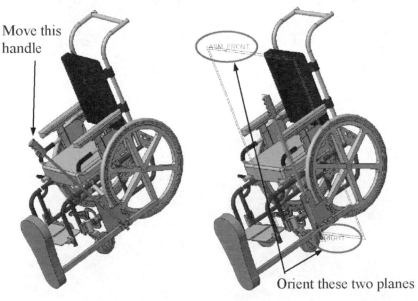

Move this handle

Orient these two planes

Figure 9-5 Align Two Axes   Figure 9-6 Coincide Two Planes   Figure 9-7

Creating a Simulation Model

As part of the simulation model, we will add a spring to connect the handle bar to the plate in order to restore the handle bar to an upright position after pulling. In addition, we will add a point force on top of the handle bar for a dynamic simulation. The force will be an impulse force of a triangular shape with a magnitude of 5 lb_f in a 0.5 second period. Enter *Mechanism* by choosing *Applications > Mechanism*.

Click the *Springs* button 🌀. From the *Graphics* window, pick *PNT0* of the handle bar and drag the handle to overlap it with *PNT0* of the plate, as shown in Figure 9-8 (you may refer to *Lesson 3* or *Lesson 8* for the detailed steps in defining a spring if necessary). In addition, enter 20 (lb_f/in.) for spring constant, and 4.5 (in.) for the free spring length. Note that the free spring length is smaller than the distance (between the two datum points) that keeps the handle bar in the upright position. Therefore, the handle bar will lean forward in the equilibrium configuration due to the spring.

Click ✅ at right to accept the definition. A spring symbol should appear in the *Graphics* window, as shown in Figure 9-8.

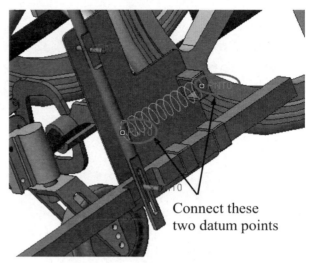

Connect these two datum points

Figure 9-8 Connecting Two Datum Points for Defining a Spring

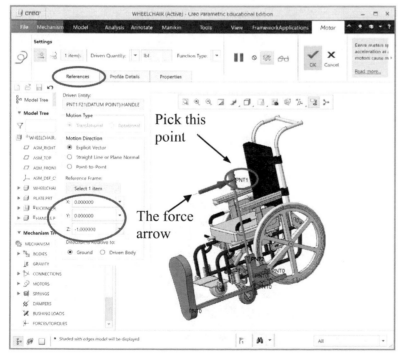

Figure 9-9 Defining Force

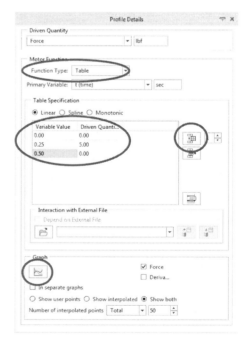

Figure 9-10 The *Profile Details* Tab

Next, we define a force following the same steps as discussed in, for example, *Lesson 3* and *Lesson 5*. Click the *Force/Torque* button ⊟ of the *Insert* group; a new set of selections will appear at the top of the *Graphics* window for defining the force (see Figure 9-9). Choose *PNT1* of the handle bar from the *Graphics* window. Click the *References* tab, and enter (0, 0, −1) for (X, Y, Z), which reverses the direction of the force, as shown in Figure 9-9. Click the *Profile Details* tab, and choose *Table* for *Function Type* under *Motor Function* (see Figure 9-10). Click the *Add rows to table* button ⊞ (circled in Figure 9-11) three times to create three empty rows. Enter three pairs of data—*0, 0*; *0.25, 5*; and *0.5, 0*—for the force magnitude. Click the *Graph* button ⊠ to show the graph of the force profile (Figure 9-11). Click the *Properties* tab and enter *Force1* for *Name*. Click the ✓ button on the right to accept the definition. A force symbol should appear at *PNT0* of the handle in the *Graphics* window, as shown in Figure 9-12.

Creating and Running a Static Analysis

We will first run a static analysis to determine the equilibrium configuration of the mechanism. This equilibrium configuration will be used as the initiation configuration for the follow-on dynamic simulation. Click the *Mechanism Analysis* button ⊠ to define an analysis. In the *Analysis Definition* dialog box appearing, enter *Static_Analysis* for *Name* and choose *Static* for *Type*. Click the *External loads* tab and remove *Force1* by clicking on the data cell and clicking the remove button ⊞ (the middle button on the right), as shown in Figure 9-13. Click *Run*. The static analysis will start and a *Graphtool* window similar to Figure 9-14 will appear showing the progress of the analysis. In the *Graphics* window, the handle bar will lean forward a bit, as shown in Figure 9-15, due to the fact that the free spring length is 4.5 in.

Again, we will use the *Drag Components* button ⊡ to create a snapshot for the current configuration, which will be used as the initial condition for dynamic simulation.

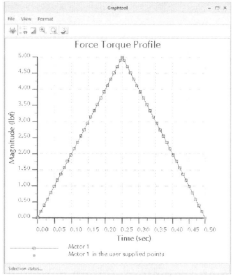

Figure 9-11 Graph of the force profile

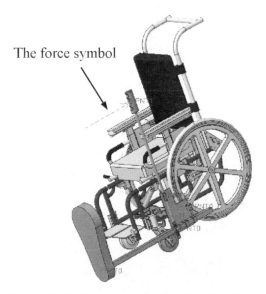

Figure 9-12 Force Symbol at the Handle Bar

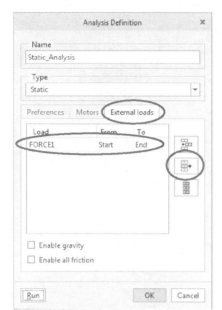

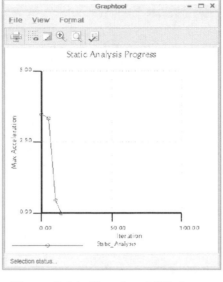

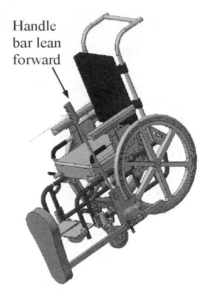

Handle
bar lean
forward

Figure 9-14 *Graphtool* Window

Figure 9-13 *Analysis*
Definition Dialog Box

Figure 9-15 Equilibrium
Configuration

Creating and Running a Dynamic Analysis

Click the *Mechanism Analysis* button [X] and enter the following in the *Analysis Definition* dialog box:

Name: *Dynamic_Analysis*
Type: *Dynamic*
Duration: *3*
Frame Rate: *100*
Minimum Interval: *0.01*
Initial Configuration: *Current*

Make sure that the external force is included this time (click the *External loads* tab of the *Analysis Definition* dialog box to confirm). Run the analysis. In the *Graphics* window, the handle bar should pull back and the kicking rod should start moving forward. After a short period (about 0.5 seconds), the motion is reversed, then back and forth until the end of the 3-second simulation period.

Saving and Reviewing Results

Click the *Playback* button [◀▶] to bring up the *Playbacks* dialog box and repeat the motion animation.

In the *Playbacks* dialog box, choose *Dynamic_Analysis* for *Result Set*, and click *Save* button [💾] to save both the results as a *.pbk* files.

We will create measures to monitor the position, velocity, and acceleration of the kicking rod along the longitudinal direction (Z-direction). All three measures will be defined at datum point *PNT0* of the foot, as shown in Figure 9-16.

Click the *Measures* button . In the *Measure Results* dialog box appearing, click the *Create New Measure* button . In the *Measure Definition* dialog box, enter *Position_Rod* for *Name*. Under *Type*, select *Position*. Pick *PNT0* in the *Graphics* window. Choose *Z-component*. Under *Evaluation Method*, leave *Each Time Step*. Click *OK* to accept the definition.

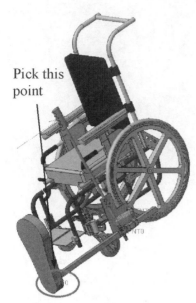

Pick this point

In the *Measure Results* dialog box, choose *Dynamic_Analysis* in the *Result Set* and click the *Graph* button on the top left corner to graph the measure. The graph should be similar to that of Figure 9-17, which shows that the kicking rod travels to about 31 in. forward due to the pulling force applied at the handle bar and about 25.5 in. backward due to the spring. The overall distance that the kicking rod travels is about 5.5 in., which probably will not produce enough momentum to kick the ball. On the other hand, the middle pin of the rod (therefore the handle bar) will collide with the two lower brackets during the backward movement. The collision is also evidenced during the motion animation. Note that the rod is moving back and forth indefinitely because no friction is being applied to any of the connections.

Figure 9-16 Defining Measures

Similarly, create measures for velocity and acceleration for the kicking rod at the same datum points along the longitudinal direction (*Z*-direction). Make sure you choose *Z-component* when you define these two measures. The graphs should look like those of Figures 9-18 and 9-19, respectively. Figure 9-18 shows that the velocity of the kicking rod is about 8 in/sec when the force is first applied. The velocity is increased to 13 in/sec when the rod is pulled back by the spring. In Figure 9-19, a maximum acceleration of 45 in/sec^2 is reached when the handle bar is released. The acceleration is increased to 70 in/sec^2 due to the stretch of the spring.

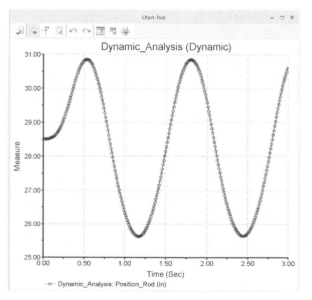

Figure 9-17 Graph of Kicking Rod Position

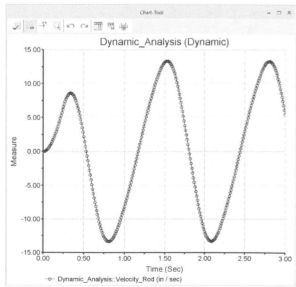

Figure 9-18 Graph of Kicking Rod Velocity

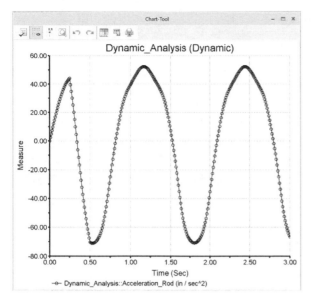

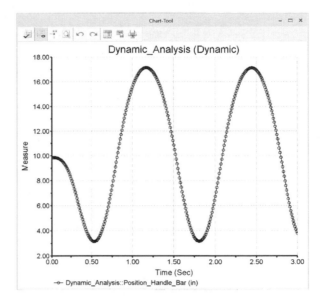

Figure 9-19 Graph of Kicking Rod Acceleration   Figure 9-20 Graph of Handle Bar Position

Next, we define the position of the handle bar in the *Z*-direction as measured at *PNT1* of the handle bar. Display the result in a graph. You should see a graph similar to that of Figure 9-20. As shown in Figure 9-20, the handle bar travels between 3 and 17 in., about 14 in. overall, which is about right for users to handle.

We completed the exercise. You may save the model by choosing *File > Save*.

9.4 Result Discussion

Looking at the results shown in the graphs, there are at least three problems revealed in the current design. First, the collision appears between the kicking rod (and the handle bar) and the lower brackets on the plate. These two brackets must move backward to provide adequate room for the rod to travel along the longitudinal direction.

Second, the spring exerts a fairly large force on the handle bar, pushing the handle bar with a larger velocity and acceleration. Since the sole purpose of the spring is to restore the handle bar to its equilibrium upward position, a large exerting force is less desirable. In order to reduce the spring force, the spring must move upward, closer to the pivot pin in order to reduce the spring deflection.

Third, the kicking rod only travels about 5.5 in., while the handle bar travels 16 in. This is due to the current position of the pivot pin. It would be more desirable to have the kicking rod travel more distance than the handle bar. This can be achieved by moving the pivot pin upward. However, by doing so, users will have to pull the handle bar with a greater force since the moment arm is reduced. There is a trade-off between the amount of the applied force and the effectiveness of the whole mechanism in terms of the kicking action. Design alternatives are to be explored, which is left as an exercise for the readers.

The design was revised by the student team. A physical device, as shown in Figure 9-21, was built by students. The physical device confirms that the contact between the kicking rod and the two brackets produces a large friction force, resulting in a large operating force to operate the device. For children with limited physical strength, such a device is unattractive. In order to reduce the friction, four bearings are added to the device, as shown in Figure 9-22. Two are added to the top surface of the kicking rod, and two are underneath it. With the bearings, the friction is significantly reduced. Therefore, a smaller force is needed to operate the device. The actual operating force is less than *20* lb$_f$.

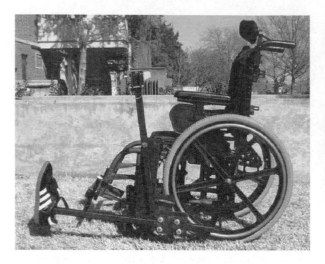
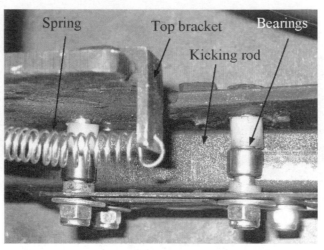

Figure 9-21 Device Assembled to the Wheelchair   Figure 9-22 Bearings Added (Top View)

Lesson 10: Kinematic Analysis for a Racecar Suspension

10.1 Overview of the Lesson

This is the second application and the final lesson of this book. In this lesson, we will take a quarter of a racecar suspension to create a motion model for kinematic analyses. The racecar model employed, as shown in Figure 10-1, is a Formula SAE (Society of Automotive Engineers) style racecar designed and built by engineering students at the University of Oklahoma (OU) during 2005-2006. Each year engineering students throughout the world design and build formula-style racecars and participate in the annual Formula SAE competitions. The result is a great experience for young engineers as a meaningful engineering project as well as an opportunity to work in a dedicated team environment.

(a) Manufactured Racecar on Display   (b) Racecar Designed in *Creo*

Figure 10-1 Formula SAE Racecar Designed and Built by OU Engineering Students

The suspension of the entire racecar was modeled for both kinematic and dynamic analyses during 2005-2006. These analysis results were validated using experimental data. The experimental data were acquired by mounting a data acquisition system on the racecar and driving the racecar on the test track following specific driving scenarios that are consistent with those of the simulations. These results were used to aid the suspension design for handling and cornering. Assembling an entire vehicle suspension for motion analysis is non-trivial and is beyond the scope of this book. Therefore, only the right front quarter of the racecar suspension, as shown in Figure 10-2, will be employed in this lesson. The purpose of this lesson is mainly to show you that *Mechanism* is capable of supporting design of kinematic characteristics of vehicle suspension, instead of repeating the detailed steps of constructing the motion model in *Mechanism*. Therefore, in this lesson, we will start with an assembled motion model. The only component we will add to the motion model is the road profile. The road profile is characterized by the geometric shape of a profile cam, which will be assembled to the tire using a cam-follower connection.

Figure 10-2 The Right Front Quarter of the Racecar Suspension (View A)

10.2 The Quarter Suspension

Physical Model

The quarter suspension consists of major components that essentially define the kinematic and dynamic characteristics of the racecar. These components include upper and lower control arms, upright, rocker, shock, push rod, tie rod, and wheel and tire, as shown in Figure 10-3. The dangling end of the shock, both control arms, rocker, and tie rod are connected to the chassis frame using numerous joints. The chassis frame is assumed fixed and the tire is pushed and pulled by the profile cam (not shown) mimicking the road profile. Two views, *View A* and *View B,* shown in Figure 10-3, are created in the assembled model and will be used for illustrations throughout this lesson.

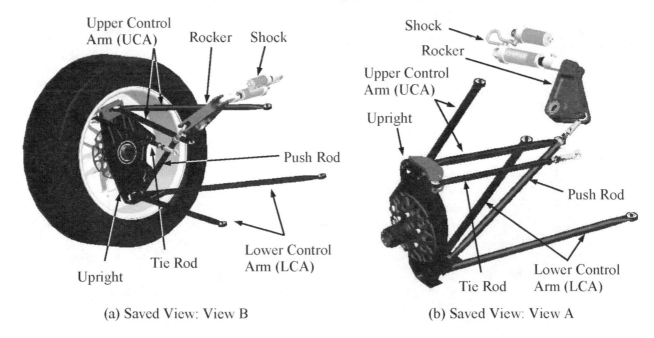

(a) Saved View: View B   (b) Saved View: View A

Figure 10-3 Major Components of the Quarter Suspension

The tire of the quarter suspension will be in contact with the profile cam representing the road profile. As shown in Figure 10-4, the geometry of the cam includes two circular arcs of radius 6.65 in. (AB and FG), which are concentric with the cam center. Therefore, when the cam rotates, these two circular arcs do not push or pull the tire, resulting in two flat segments of the road profile, as shown in Figure 10-5. In addition, the circular arc CDE is centered 4 in. above the cam center with a radius of 4 in. Therefore, when the cam rotates, arc CDE pushes the tire up, mimicking a hump of 1.35 in. (that is 1.35 = 8–6.65, peak at point D). A ditch is characterized by an 8 in. arc (HIJ) centered at 3 in. above the cam center. As the cam rotates, arc HIJ creates a ditch of 1.65 in. deep [that is 1.65 = 6.65–(8–3)]. The remaining straight lines and arcs provide smooth transitions among flats, humps, and ditches in the road profile.

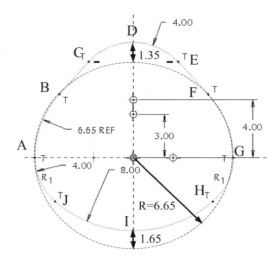

Figure 10-4 Geometry of the Profile Cam

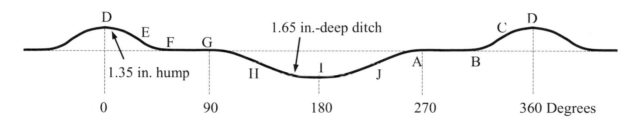

Figure 10-5 Road Profile Generated by the Profile Cam

Based on the geometry of the profile cam, this quarter suspension will go over a 1.35 in. hump and a 1.65 in. ditch in one complete rotation of the profile cam. Note that since the radius of arc AB is 6.65 in., the cam will cause the quarter suspension to travel roughly 41.8 in. (3.48 ft.) in one complete rotation. Since the profile cam will rotate a complete cycle in one second, the suspension travels about 3.48 ft/sec, i.e., 2.37 MPH, a very slow motion.

Creo Parts and Assembly

All *Creo* parts and assemblies are provided for this lesson. In addition, all parts and subassemblies are assembled, except for the profile cam (part name: *profile.prt*). The profile cam will be assembled to the tire using a cam-follower connection. A servo motor will be added to drive the cam at a constant angular velocity of 360 degrees/sec, therefore pushing and pulling the tire along the vertical direction, mimicking the situation where the racecar goes over humps and ditches.

Note that the example files you downloaded from the publisher's website should consist of 51 files: 38 parts, 12 assemblies, and 1 result, as listed in Table 10-1. The quarter suspension *quarter_suspension.asm* is completely assembled except for the road profile (*profile.prt*). We will start with this assembly and bring in the profile cam, which will be assembled to the tire using a cam-follower connection. In addition, a completely assembled motion model, *motion_quarter_suspension.asm*, is included for your reference. A simulation result file, *AnalysisDefinition1.pbk*, is also included. You may want to open this motion model and bring up this result file to see the motion animation of the quarter suspension system.

Table 10-1 List of Files in Lesson 10 Folder

Assembly	Part/Subassemblies			Remarks
quarter_suspension.asm				Assembly to start the lesson
	hardpoints.prt			Datum features
	fr_rocker.asm			Rocker
		susp_front_bellcrank.prt		
		susp_rocker_bearing_61903_2RS1.prt (2)		
	shock_upper.prt			Shocks
	shock_lower.prt			
	lcm.asm			Lower Control Arm
		susp_front_rh_lower_a_am.prt		
		susp_025_hab_4t_special_race.prt (2)		
		susp_a_arm_flare.prt (2)		
		susp_3125_pwb_5tg_circ_race.prt		
	prod.asm			Push Rod
		susp_front_push_rod.prt		
		susp_025_pr_com_4_race.prt		
		bolt_03125_allthread_rod_link.prt		
		susp_025_fem_rod_end_asw_4t.prt		
		susp_pushrod_flare.prt		
	fr_upright.asm			Upright
		susp_front_rh_upright.prt		
		susp_front_camber_shim.asm		
			susp_front_camber_shim_8.prt (2)	
			susp_front_camber_shim_8.prt (4)	
		susp_front_rh_steer_arm.asm		
			susp_front_rh_steer_arm.prt	
			susp_front_steer_ackerman_plate.prt	
	lcm.asm			Lower Control Arm
		susp_front_rh_upper_a_am.prt		
		susp_025_hab_4t_special_race.prt (2)		
		susp_a_arm_flare.prt (4)		
		susp_025_hab_4t_special_race2.prt (2)		
	wheel.asm			Wheel
		susp_front_hub.prt		
		susp_wheel_drive_pin.prt (4)		
		susp_front_hub_spacer.prt		
		susp_front_wheelbear_61908_2RS1.prt (2)		
		brake_rotor_hat.prt		
		susp_wheel_assembly.asm		
			susp_wheel_inner_rim.prt	
			susp_wheel_outer_rim.prt	
			susp_wheel_center.prt	
			susp_tire_13inch_20_x_6.prt	
	trod.asm			Tie Rod Assembly
		susp_steering_front_tie_rod.prt		
		susp_025_fem_rod_end_asw_4t.prt		
		Bolt_3125_allthread_rod_link.prt		
		susp_025_hab_4t_special_race.prt		
		susp_front_tie_rod_flare.prt		
	profile.prt			To be assembled
motion_quarter_suspension.asm				Complete Motion Model
AnalysisDefinition1.pbk				Result Data File

Note that the in-lb$_f$-sec unit system has been chosen for all parts and assemblies. You may want to choose (from the pull-down menu) *File > Prepare > Model Properties* to ensure the choice of the proper unit system.

Motion Model

There are nine bodies in this motion model, including the ground body. All the key datum features, including datum coordinate systems, datum axes, and datum points, required for assembly are collected in the part called *hardpoints.prt*, as shown in Figure 10-6a. The *hardpoint.prt* was assembled to the overall assembly *quarter_suspension.asm* by aligning coordinate systems *GCS* (*hardpoints.prt*) and *ASM_DEF_CSYS* (*quarter_suspension.asm*). Therefore, *hardpoints.prt* belongs to the ground body, and the coordinate system, *GCS* or *ASM_DEF_CSYS*, becomes the *WCS* (World Coordinate System). Note that the *X*-direction is the forward direction, as shown in Figure 10-6a, and the ground is about 1.4 in. above *WCS* along the *Z*-direction (that is the distance between the lowest point in the tire and the coordinate system *GCS*).

There are two rigid (no symbol), three pin, eight ball, and one cylinder joints, as shown in Figure 10-6b and listed in Table 10-2. Note that axis *A_1* in *hardpoints.prt* is used for assembling the profile cam later. A kinematic analysis will be created with a servo motor that rotates the profile cam for 2 seconds. Three measures will be defined to monitor the characteristics of the suspension, including vertical wheel travel, shock travel, and camber angle.

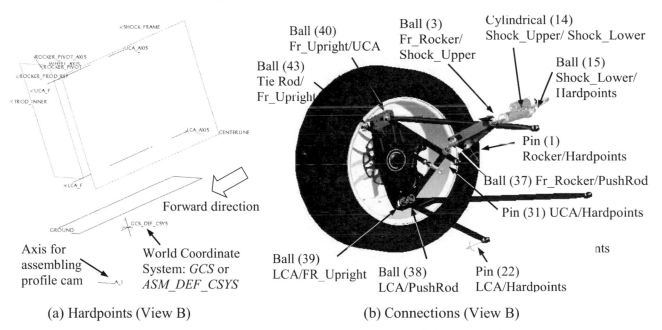

(a) Hardpoints (View B)   (b) Connections (View B)

Figure 10-6 Joints Defined for the Quarter Suspension Assembly

10.3 Using *Mechanism*

Assembling the Profile Cam

Start *Creo*, select working directory, and open *quarter_suspension.asm*. You should see *quarter_suspension.asm* appear in the *Graphics* window with default view *View B* (see Figure 10-3a). Note that the *quarter_suspension.asm* consists of ten components, listed in the model tree window, i.e., *HARDPOINTS.PRT*, *FR_ROCKER.ASM*, *SHOCK_UPPER.PRT*, *SHOCK_LOWER.PRT*, *LCA.ASM*,

PROD.ASM, FR_UPRIGHT.ASM, UCA.ASM, WHEEL.ASM, and *TROD.ASM.* You can also see more details of the bodies and connections in the motion model by expanding the motion entities in the model tree (lower half) and right clicking them to choose *Edit Definition.* You may choose *Applications > Mechanism* to enter the *Mechanism.* The joint symbols appear, similar to those of Figure 10-6b. The *Model Tree* window is split into two sections. You can find the body and connection definitions by expanding these entities in the lower section, as shown in Figure 10-7. Note that rigid connections are not listed in the model tree.

Table 10-2 Connections and Placement Constraints

Body	Part or Assembly	Connections	Assembled to	Placement Constraints
Ground Body	*hardpoints.prt*	Rigid	*quarter_ suspension.asm*	Coordinate Systems Alignment: *GCS* to *ASM_DEF_CSYS*
Body1	*ft_rocker.asm*	Pin (1)	*hardpoints.prt*	Axis Alignment: *ROCKER_PIVOT_AXIS* to *ROCKER_PIVOT_AXIS* Translation: *ROCKER_PIVOT_CENTER* to *ROCKER_PIVOT*
Body2	*shock_upper.prt*	Ball (3)	*hardpoints.prt*	Point Alignment: *SHOCK_UPPER* to *FR_ROCKER*
Body3	*shock_lower.prt*	Ball (15)	*hardpoints.prt*	Point Alignment: *SHOCK_LOWER* to *SHOCK_FRAME*
		Cylinder (14)	*shock_upper.prt*	Axis Alignment: *A_3* to *A_1*
Body4	*lca.asm*	Pin (22)	*hardpoints.prt*	Axis Alignment: *LCA_AXIS* to *LCA_AXIS* Translation: *LCA_FR* to *LCA_F*
Body5	*prod.asm*	Ball (37)	*lca.asm*	Point Alignment: *PROD_INNER* to *ROCKER_PROD*
		Ball (38)	*fr_rocker.asm*	Point Alignment: *PROD_OUTER* to *LCA_PROD*
Body6	*uca.asm*	Pin (31)	*hardpoints.prt*	Axis Alignment: *UCA_AXIS* to *UCA_AXIS* Translation: *UCA_FR* to *UCA_F*
		Ball (40)	*upright.prt*	Point Alignment: *UCA_OUTER* to *UPRIGHT_UCA*
Body7	*fr_upright.asm*	Ball (39)	*lca.asm*	Point Alignment: *UPRIGHT_LCA* to *LCA_OUTER*
	wheel.asm	Rigid (42)	*fr_upright.asm*	Axis Alignment: *A_1* to *A_1* Surface Mate: *F2* to *F9*
Body8	*trod.asm*	Ball (43)	*fr_upright.asm*	Point Alignment: *TIEROD_OUTER* to *TROD_UPRIGHT*
		Ball (44)	*hardpoints.prt*	Point Alignment: *TIEROD_INNER* to *TROD_INNER*

Next, we bring in the only component, *profile.prt*, to the assembly. Click the *Assemble* button and choose *profile.prt.* The profile cam will be brought in with a default position and orientation similar to those of Figure 10-8. Note that you may need to zoom out to see the profile cam. Move the part closer to the quarter suspension. We will define a pin joint to assemble the cam (*profile.prt*) to the ground body (*hardpoints.prt*). This pin joint will be defined by aligning axis *A_1* in *profile.prt* to *A_1* in *hardpoints.prt* and aligning datum plane *FRONT* in *profile.prt* to *ASM_TOP* in the quarter suspension assembly.

Turn on the datum axis display. You should see a number of axes appear, including the two axes we need. In the *Component Placement* dashboard from the *User Defined* list, choose the *Pin* joint. Turn on datum axis display and pick *A_1* (*profile.prt*) and *A_1* (*hardpoints.prt*), as shown in Figure 10-8. After picking the axes, turn off the datum axis display and turn on datum plane display. Pick *FRONT* datum plane in *profile.prt* to *ASM_TOP* in *quarter_suspension.asm*. Note that you may need to zoom out to see datum plane *ASM_TOP*. The profile cam is not properly positioned since the two datum planes are not aligned. We will position *profile.prt* by choosing *Distance* constraint and entering an offset of 24.5 using the *Placement* window, as shown in Figure 10-9.

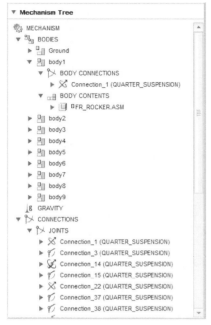

Figure 10-7 Bodies and Joints Listed in the *Model Tree* Window

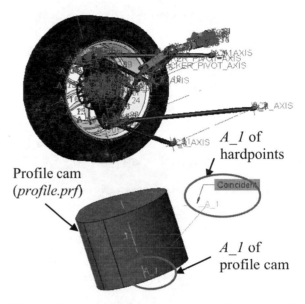

Figure 10-8 *profile.prt* Brought in for Assembly

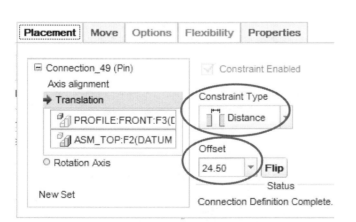

Figure 10-9 Entering Offset for Plane Alignment

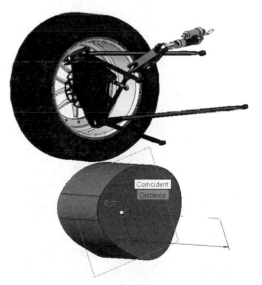

Figure 10-10 Defining a Pin Joint for the Profile Cam (*profile.prt*)

The pin joint is now fully defined, as indicated by the status message in the *Placement* dashboard. However, the part may not be properly oriented, as shown in Figure 10-10. We want to have the hump of the profile cam at the bottom, i.e., the suspension being pulled down. We will use the *Drag Components* button  later to create a snapshot of the desired configuration for initial condition. For now, accept the definition by clicking the ☑ button.

Creating a Simulation Model

Enter *Mechanism* by choosing *Applications > Mechanism*.

All connection symbols appear similar to those of Figure 10-6b. We will choose the cylindrical surfaces of the profile cam and the tire to define the cam-follower.

Click the *Cam-Follower* button ⬚ of the *Connections* group at the top of the *Graphics* window. The *Cam-Follower Connection Definition* dialog box appears (Figure 10-11). Use the default name (*CamFollower1*), choose *Autoselect*, and click the *Select* button ⬚. Pick the cylindrical surface of the profile cam; all the surrounding surfaces will be selected (see Figure 10-12). Click the *OK* button in the *Select* dialog box (right underneath the *Cam-Follower Connection Definition* dialog box). The surface selected will appear in the *Surfaces/Curves* text area. An arrow will also appear on the surfaces showing the surface normal (pointing in outward direction).

Figure 10-11

Click the *Cam2* tab and repeat the same process by selecting the outer cylindrical surface of the tire, as shown in Figure 10-12. Click *OK* in the *Cam-Follower Connection Definition* dialog box, a cam-follower symbol ⬚ will appear between the two surfaces selected, and the profile and the tire will be oriented so that their outer surfaces are in contact.

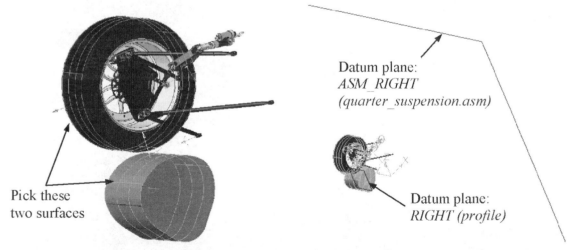

Pick these two surfaces

Datum plane:
ASM_RIGHT
(quarter_suspension.asm)

Datum plane:
RIGHT (profile)

Figure 10-12 Creating Cam-Follower Connection   Figure 10-13 Align Two Datum Planes

Click the *Drag Components* button  at the top of the *Graphics* window, click the profile cam, and rotate it by moving the mouse. You should see that the tire is being pushed up and pulled down, and all the suspension components are moving accordingly. Rotate the profile cam to allow the tire to be its lowest position possible. Create a snapshot by orienting two datum planes, e.g., *RIGHT* of *profile.prt* and *ASM_RIGHT* of *quarter_suspension.asm*, as shown in Figure 10-13 (you may want to zoom out the view to see *ASM_RIGHT*). Save the snapshot. Now the assembly is completed.

Next, we create a servo motor at the pin joint between the profile cam and the ground body. The servo motor will rotate the profile cam with a constant angular velocity of 360 degrees/sec; therefore, pushing and pulling the tire, mimicking humps and ditches.

Click *Servo Motors* button of the *Insert* group on top of the *Graphics* window; a new set of selections will appear (see Figure 10-14). Pick the pin joint of the profile cam from the *Graphics* window. The rotation axis must point in a direction like that of Figure 10-14. Click the *References* tab to make sure that *Connection_49.first_rot_axis* has been selected for *Driven Entity*, as circled in Figure 10-14. Next, click the *Profile Details* tab, choose *Angular Velocity* for *Driven Quantity*, and leave *Constant* (default) for *Function Type* under *Motion Function*. Enter *360* for the constant *A* (360 degrees/sec, i.e., 60 rpm).

Click the *Properties* tab to enter *Motor1* for name. Click the button on the right to accept the definition. A motor symbol appears in the *Graphics* window overlapping with the pin joint (Figure 10-16).

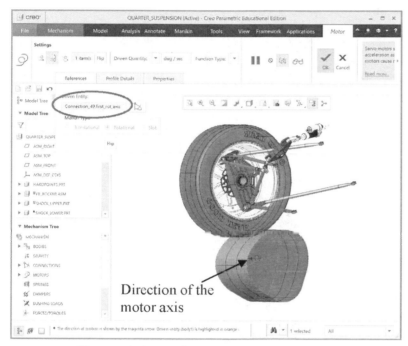

Figure 10-14 Defining a Servo Motor

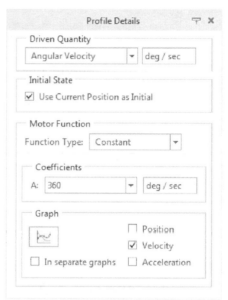

Figure 10-15

Creating and Running a Kinematic Analysis

Click the *Mechanism Analysis* shortcut button and enter the following in the *Analysis Definition* dialog box appearing:

Name: *Kinematic_Analysis*
Type: *Kinematic*
Start Time: *0*
End Time: *2*
Frame Rate: *100*
Minimum Interval: *0.01*
Initial Configuration: *Current*

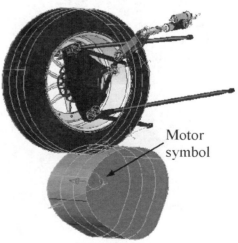

Figure 10-16 Motor Added to Pin Joint in Profile Cam

Click the *Motors* tab to make sure that *Motor1* is listed and is included in this analysis. Run the analysis. In the *Graphics* window, the profile cam should start turning, pushing and pulling the tire vertically, causing the suspension components to move. The profile cam should make two complete turns.

Saving and Reviewing Results

Click the *Playback* button ◀▶ of the *Analysis* group on top of the *Graphics* window to bring up the *Playbacks* dialog box and repeat the motion animation. On the *Playbacks* dialog box, click the *Save* button 🖫 to save the results as a *.pbk* file.

As mentioned earlier, we will create three measures to monitor the characteristics of the suspension. Note that these three measures provide initial understanding of the suspension design. They do not necessarily tell all the details regarding the pros and cons of the design. These measures are the vertical wheel travel, shock travel, and camber angle. Note that the camber angle will be defined as the rotation of the upright along the *X*-axis of *WCS*.

Click the *Measures* button ⊠ of the *Analysis* group on top of the *Graphics* window. In the *Measure Results* dialog box, click the *Create New Measure* button ▢. In the *Measure Definition* dialog box, enter *Vertical_Wheel_Travel* for *Name* (Figure 10-17). Under *Type*, select *Position*. Pick datum point *PNT20* in *susp_wheel_center.prt* for *Point or motion axis* (see Figure 10-18), and *GCS* of *hardpoints.prt* (for *Coordinate system*). Choose *Z-component* for *Component*. Under *Evaluation Method*, leave *Each Time Step*. Click *OK* to accept the definition.

Figure 10-17

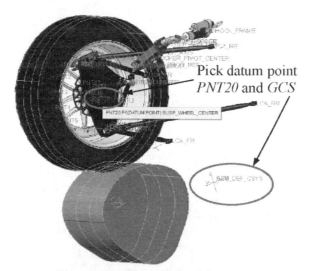

Figure 10-18 Defining Measure: *Vertical_Wheel_Travel*

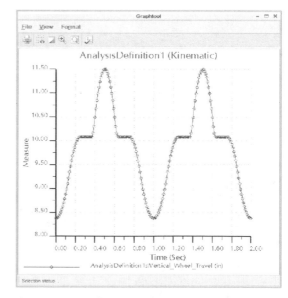

Figure 10-19 Graph of *Vertical_Wheel_Travel*

In the *Measure Results* dialog box, choose *AnalysisDefinition1* in the *Result Set* and click the *Graph* button ⬚ at the top left corner to graph the measure. The graph should be similar to that of Figure 10-19, which shows the vertical position of the center point of the wheel center. This measure reflects the road profile that the wheel travels. Note that the data do not exactly depict the road profile due to the tire camber angle. The center of the wheel travels vertically between about 11.5 in. and 8.4 in. The flat portion (10.1 in.) reassembles the flat road profile. The distance between the peak and the flat portion is about 1.4 in. due to the 1.35 in. hump. Similarly, the distance between the flat portion and the crest is about 1.7 in. due to the 1.65 in. ditch.

The second measure is the shock travel distance. In the *Measure Definition* dialog box, enter *Shock_Travel* for *Name*. Under *Type*, select *Separation* (see Figure 10-20). Pick two datum points: *SHOCK_FRAME* of *hardpoints.prt* and *ROCKER_SHOCK* of *fr_rocker.asm* (see Figure 10-21) in the *Graphics* window. You may click the right mouse button to bring out *ROCKER_SHOCK* from the overlapped datum points. Choose *Distance* for *Separation Type*. Under *Evaluation Method*, leave *Each Time Step*. Click *OK* to accept the definition.

Follow the same steps to display the graph. The graph should be similar to that of Figure 10-22, which shows that the shock travels between about 6 in. and 8.5 in. The overall travel distance is about 2.5 in., which is probably too large for such a small hump or ditch. In fact, in the simulation, it appears that the shock is compressed too much, in which the piston penetrates into its reserve cylinder. In reality, this will not happen. However, the simulation raises a flag indicating that there could be severe contact within the shock, leading to potential part failure.

Figurc 10-20

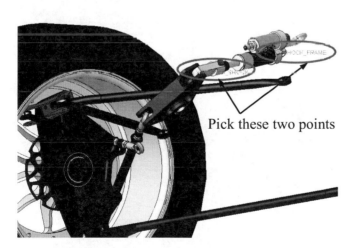

Figure 10-21 Defining Measure: *Shock_Travel*

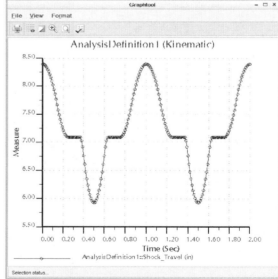

Figure 10-22 Graph of *Shock Travel*

The third measure is the camber angle. The camber angle is the angle made by the wheel of an automobile; specifically, it is the angle between the vertical axis of the wheel and the vertical axis of the vehicle when viewed from the front or rear. Camber angle is an important measure that contributes to the performance of steering and suspension. If the top of the wheel is further out than the bottom (that is, away from the axle), it is called positive camber; if the bottom of the wheel is further out than the top, it is called negative camber. In this model, the camber angle will be defined as the rotation angle of the upright along the *X*-axis of *GCS*. For this measure, enter *Camber_Angle* for *Name* (see Figure 10-23).

Under *Type*, select *Body*, and pick *body7* (*fr_upright.asm*) in the *Mechanism* model tree window (lower half, see Figure 10-7). Choose *Orientation* for *Property*, and click *1* for *Euler Component*, which means the rotation measure is defined along the *X*-axis (of the *WCS* as default). Leave *WCS* as the *Coordinate System* (default), which is *GCS* or *ASM_CSYS_DEF*. Under *Evaluation Method*, leave *Each Time Step*. Click *OK* to accept the definition.

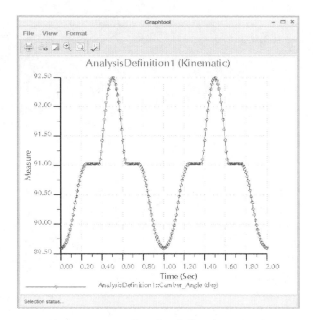

Figure 10-23   Figure 10-24   Graph of *Camber Angle*

As shown in Figure 10-24, the camber angle was set to be about 91 degrees on the flat terrain. The camber angle varies to 92.5 and 89.5 degrees, respectively, when the tire goes over the hump and the ditch. In general, camber angle alters the handling quality of a particular suspension design. In general, negative camber improves grip when cornering. This is because it places the tire at a more optimal angle to the road, transmitting the forces through the vertical plane of the tire, rather than through a shear force across it. However, excessive negative camber change in hump can cause early lockup under breaking or wheel spin under acceleration. There is only limited information that can be obtained by conducting kinematic analysis of the quarter suspension. Ultimately, a full-vehicle dynamic simulation must be carried out to fully understand the suspension design and hopefully develop a strategy for design improvement.

10.4 Full-Vehicle Dynamic Simulation

A final note: a full-vehicle dynamic simulation model was created in *ADAM/Car*, using the model templates provided, as shown in Figure 10-25a. With *ADAMS/Car*, users can simply enter vehicle model data into the templates, and *ADAMS/Car* will automatically construct subsystem models, such as engine, shock absorbers, tires, as well as the full vehicle assemblies. Once these templates are created, they can be made available to novice users, enabling them to perform standardized vehicle maneuvers. The vehicle model was then simulated for various test scenarios, including skid pad racing, which is a constant radius cornering simulation, as shown in Figure 10-25b.

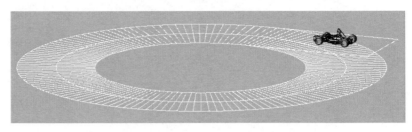

(a) A 15 dofs *ADAMS/Car* Model   (b) Skid Pad Racing Simulation

Figure 10-25 Vehicle Dynamic Simulation of Formula SAE Racecar

Notes:

APPENDIX A: DEFINING JOINTS

Degrees of Freedom

Understanding degrees of freedom is critical in selecting appropriate connections or joints to create a motion simulation model. In mechanical systems, the number of degrees of freedom (dofs) represents the number of independent parameters required to specify the position, velocity, and acceleration of individual moving bodies in the system. A completely unconstrained body has six degrees of freedom: three translational and three rotational. If you apply a pin connection (or joint) to the body, you restrict its movement to rotation about an axis, and the degrees of freedom for the body are reduced from six to one.

In most mechanical systems, you can determine the degrees of freedom using the following formula:

$$D = 6M - N \hspace{8cm} \text{(A-1)}$$

where D is the degrees of freedom of the mechanism, M is number of bodies not including the ground body, and N is the number of degrees of freedom restrained by all connections.

The number of bodies can be found from the model summary window. You may bring up this window by selecting *Analysis > Mechanism*, and then click the *Summary* button ⋈ of the *Information* group on top of the *Graphics* window.

You may apply Eq. A-1 to a door model that is supported by two hinges using pin joints. Note that a pin joint imposes 5 dofs. Insert the number of bodies and dofs into this equation:

$$D = (6 \times 1) - (2 \times 5) = -4.$$

The calculated degrees of freedom result is −4, which is unrealistic.

Mechanisms should not have negative degrees of freedom. For most cases, we would like to have the number of dofs equal 1. The challenge is to choose the joints that will give you one dof and still allow the intended motion.

The equations above do not account for the influence of the drivers (or servo motors) in your mechanism. Servo motors are defined to drive joint displacements (or rotation), velocities, or accelerations. Drivers will eliminate remaining degrees of freedom.

Redundancy

Redundancies are excessive dofs. When a joint constrains the model in exactly the same way as another joint (like the door hinge example), the model contains excessive dofs, also known as redundancies. A joint becomes excessive when it does not introduce any further restriction on a body's motion.

It is important that you eliminate redundancies as much as you can from your model while carrying out dynamic analyses. If you do not remove redundancies, you may not get accurate values when you measure connection reactions or load reactions.

For example, if you model a door using two pin joints for the hinges, the second pin joint does not contribute to constraining the door's motion. The software detects the redundancies and ignores one of the pin joints in its analysis. The outcome may contain incorrect reaction results, yet the motion is correct.

For complete and accurate reaction forces, it is critical that you eliminate redundancies from your mechanism whenever possible.

For strictly kinematic problems where you are interested in displacement, velocity, and acceleration, redundancies in your model do not alter the design and performance of the mechanism.

You can control the redundancies in your model by your choice of connections. These joints must be able to restrict the same dofs but not duplicate each other. After you decide which connections you want to use, you can use Eq. A-1 to calculate the dofs and check redundancies.

By default, the software calculates the dofs and redundancies for the model each time you analyze its motion. To check if your model has redundancies, first run a dynamic, static, or force balance analysis. Use the *Measure Results* dialog box to calculate the dofs and redundancies in your mechanism, as described next.

Click *Measures* button ⊠ of the *Analysis* group on top of the *Graphics* window. In the *Measure Results* dialog box (Figure A-1), click the *Create New Measure* button ▢. In the *Measure Definition* dialog box (Figure A-2), enter *Degrees_of_Freedom* for *Name* (or any other name you prefer). Under *Type*, select *System*. Choose *Degrees of Freedom* for *Property*. Click *OK* to accept the definition. Note that if you have already run an analysis you should see the degrees of freedom displayed in the window (Figure A-1).

Figure A-2 The *Measure Definition* Dialog Box

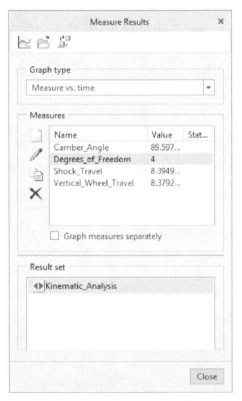

Figure A-1 The *Measure Results* Dialog Box

Joint Types in *Mechanism*

Before you select a joint to apply to your model, you should know what movement you want to restrain for the body and what movement you want to allow. The following table describes the common joint types you can choose to create motion models and their corresponding free degrees of freedom.

Joint Type	Number of Dofs			Remarks
	Rotation	Translation	Total	
Weld	0	0	0	Glues two bodies together.
Rigid	0	0	0	Glues two parts together while changing the underlying body definition. Parts constrained by a rigid connection constitute a single body.
Slider	0	1	1	Translates along an axis Plane-plane coincident
Pin	1	0	1	Rotates about an axis
Cylinder	1	1	2	Translates along and rotates about a specific axis Point on line Plane–plane orient
Ball	3	0	3	Rotates in any direction Point–point align
Planar	1	2	3	Bodies connected by a planar joint move in a plane with respect to each other. Rotation is about an axis perpendicular to the plane. Plane–plane align/mate
Bearing	3	1	4	Combines a ball joint and a slider joint Point on line
Slot	3	1	4	Point on a non-straight trajectory
Gimbal	3	0	3	Aligns the centers of two CSYS.

Notes:

APPENDIX B: DEFINING MEASURES

Types of Measure

When you click the *New* button ☐ on the *Measure Results* dialog box (Figure B-1), the *Measure Definition* dialog box opens (Figure B-2). You can create measures for specific model entities or for the entire mechanism. You can also include measures in your own expressions for user-defined measures.

You can create the following measures in *Mechanism*.

- Position—Measure the location of a point, vertex, or motion axis during the analysis.
- Velocity—Measure the velocity of a point, vertex, or motion axis during the analysis.
- Acceleration—Measure the acceleration of a point, vertex, or motion axis during the analysis.
- Connection Reaction—Measure the reaction forces and moments at joint, gear-pair, cam-follower, or slot-follower connections.
- Net Load—Measure the magnitude of a force load on a spring, damper, servo motor, force, torque, or motion axis. You can also confirm the force load on a force motor.
- Loadcell Reaction—Measure the load on a loadcell lock during a force balance analysis.
- Impact—Determine whether impact occurred during an analysis at a joint limit, slot end, or between two cams.
- Impulse—Measure the change in momentum resulting from an impact event. You can measure impulses for joints with limits, for cam-follower connections with liftoff, or for slot-follower connections.
- System—Measure several quantities that describe the behavior of the entire system.
- Body—Measure several quantities that describe the behavior of a selected body.
- Separation—Measure the separation distance, separation speed, and change in separation speed between two selected points.
- Cam—Measure the curvature, pressure angle, and slip velocity for either of the cams in a cam-follower connection.
- User Defined—Define a measure as a mathematical expression that includes measures, constants, arithmetical operators, *Creo* parameters and algebraic functions.

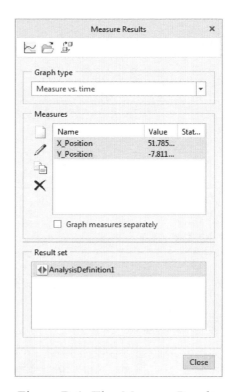

Figure B-1 The *Measure Results* Dialog Box

Figure B-2 The *Measure Definition* Dialog Box

About Measures Associated with Model Entities

This table organizes *Mechanism* measures according to the type of model entity that you select to define the measure.

Entity	Measure
Point	Position, Velocity, Acceleration, Separation—distance, speed, change in speed
Motion axis	Position, Velocity, Acceleration, Net load
Joint connection	Connection reaction, Impact, Impulse
Cam-follower connection	Cam—curvature, pressure angle, slip velocity, Connection reaction, Impact, Impulse
Slot-follower connection	Connection reaction, Impact, Impulse
Gear-pair connection	Connection reaction
Spring, damper, force, torque, servo motor, force motor	Net load

About Measure Results

Measures can help you understand and analyze the results of moving a mechanism and provide information to improve the mechanism's design. Before you can calculate and view measure results, you must have run or saved and restored results from one or more analyses for your mechanism.

You may find more information about the measures that individual analysis produces by reviewing help (click the *Help* button ❓ at the top right corner of the *Creo* window) by choosing, as shown in Figure B-3,

Simulation > Mechanism Design and Mechanism Dynamics > Mechanism Design > Creating Mechanism Models > Measures, Graphs, and Evaluation Methods > Measures > About Measure Results

Or directly enter from the *Browser*:

https://support.ptc.com/help/creo/creo_pma/r6.0/usascii/index.html#page/simulate%2Fmech_des%2Fmeasures%2Fmeas_evaln_methods.html%23

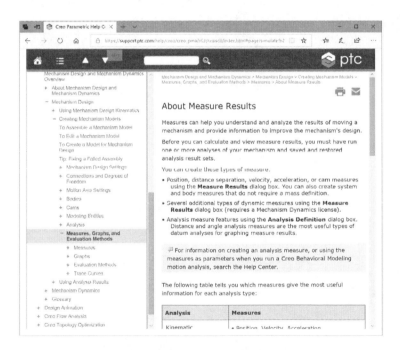

Figure B-3 The *Help* Window: About Measure Results

About Evaluation Methods

When you define dynamics measures, you can choose from several evaluation methods. The graph of the measure and the quantity displayed under *Value* on the *Measure Results* dialog box are different for different evaluation methods.

You may find more information about the measures that individual analysis produces by reviewing help (click the Help button ❓ at the top right corner of the *Creo* window) by choosing, as shown in Figure B-4,

Simulation > Mechanism Design and Mechanism Dynamics > Mechanism Design > Creating Mechanism Models > Measures, Graphs, and Evaluation Methods > Evaluation Methods > About Evaluation Methods

Or directly enter from the *Browser*:

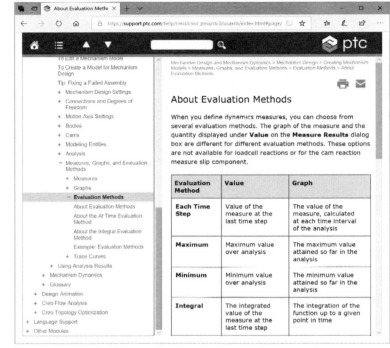

Figure B-4 The *Help* Window: About Evaluation Methods

http://help.ptc.com/creo_hc/creo30_pma_hc/usascii/index.html#page/simulate%2Fmech_des%2Fmeasures%2Fmeas_evaln_methods.html

Notes:

APPENDIX C: THE DEFAULT UNIT SYSTEM

The *in-lb$_m$-sec* Unit System

The default unit system employed by *Creo*, therefore the *Mechanism*, is *in-lb$_m$-sec* (inch-pound mass-second). This unit system is not quite common to many engineers. The basic physical quantities involved in determining a unit system are length, time, mass, and force. These four basic quantities are related through Newton's second law,

$$F = ma \qquad\qquad (C-1)$$

where *F*, *m*, and *a* are force, mass, and acceleration (length per second square), respectively.

In the default unit system, *in-lb$_m$-sec*, the force unit will be determined by length (in.), mass (lb$_m$), and second (sec) through Eq. C-1; i.e.,

$$1\ lb_m\ in/sec^2\ (force) = 1\ lb_m\ (mass) \times 1\ in/sec^2\ (acceleration) \qquad\qquad (C-2)$$

where the force unit, lb$_m$ in/sec^2, is a derived unit.

From Eq. C-2, a 1 lb$_m$ in/sec^2 force will generate a 1 in/sec^2 acceleration when applied to a 1 lb$_m$ mass block, as shown in Figure C-1a. The same block will weigh 1 lb$_f$ on earth (see Figure C-1b), where the gravitational acceleration is assumed 386 in/sec^2; i.e.

$$1\ lb_f\ (force) = 1\ lb_m\ (mass) \times 386\ in/sec^2\ (acceleration) \qquad\qquad (C-3)$$

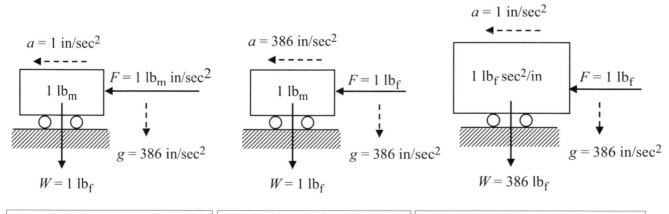

(a) A 1 lb$_m$ in/sec^2 Force Applied to a 1 lb$_m$ Mass Block

(b) A 1 lb$_f$ Force Applied to a 1 lb$_m$ Mass Block

(c) A 1 lb$_f$ Force Applied to a 1 lb$_f$ sec^2/in Mass Block

Figure C-1 Forces Applied on Blocks of Different Masses

Therefore, from Eqs. C-2 and C-3, we have 1 lb_f = 386 lb_m in/sec^2. That is, the force quantity entered into *Mechanism* is in the lb_m in/sec^2 unit by default, which is 386 times smaller than the 1 lb_f that we are more used to. When you apply a 1 lb_f force to the same mass block, it will accelerate 386 in/sec^2, as shown in Figure C-1c. Therefore, you must be very careful in entering numerical figures while defining your analysis models. For example, if you apply a 1,000 unit force to a mechanical component in the default unit system, it is indeed just a 1,000/386 = 2.59 lb_f, a very small force.

On the other hand, we have the mass unit 1 lb_m = 1/386 lb_f sec^2/in. It means that a 1 lb_m mass block is 386 times smaller than that of a 1 lb_f sec^2/in block. Therefore, a 1 lb_f sec^2/in block will weigh 386 lb_f on earth. When applying a 1 lb_f force to the mass block, it will accelerate at a 1 in/sec^2 rate, as illustrated in Figure C-1c.

APPENDIX D: FUNCTIONS

Depending on the type of motion you want to impose on your mechanism, you can choose a function to define the magnitude of your servo motors or force motors. The table in the *Creo Help* (see Figure D-1) lists different types of functions that are used to generate the magnitude. You need to enter the values of the coefficients for the functions. The value of x in the function expressions is supplied by the simulation time or, for force motors, by either the simulation time or a measure you select.

You can find this table by clicking the *Help* button ❓ at the top right corner of the *Creo* window and choosing

Simulation > Mechanism Design and Mechanism Dynamics > Mechanism Design > Creating Mechanism Models > Modeling Entities > Servo Motors > To Define a Motor

Or directly enter from the *Browser*:

http://support.ptc.com/help/creo /creo_pma/usascii/index.html#p age/simulate%2Fmech_des%2F motors%2FTo_Define_a_Motor .html%23wwconnect_header

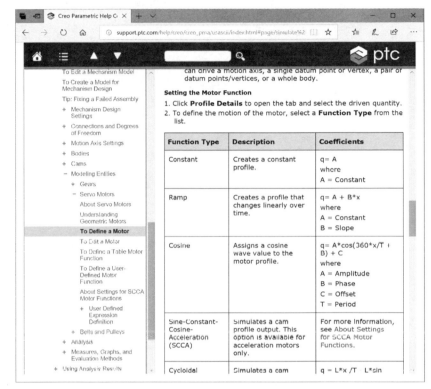

Figure D-1 The *Creo Help* Window: Setting the Motor Function

Notes: